WHY DOES MATH WORK ... IF IT'S NOT REAL?

According to G. H. Hardy, the "real" mathematics of the greats, such as Fermat and Euler, is "useless," and thus the work of mathematicians should not be judged on its applicability to real-world problems. Yet, mysteriously, much of mathematics used in modern science and technology was derived from this "useless" mathematics. Cell phone technology is based on trig functions, which were invented centuries ago. Newton observed that the earth's orbit is an ellipse, a curve discovered by ancient Greeks in their futile attempt to double the cube. It is as if some magic hand had guided the ancient mathematicians, so their formulas were perfectly fitted for the sophisticated technology of today. Using anecdotes and witty storytelling, this book explores that mystery. Through a series of fascinating stories of mathematical effectiveness, including Planck's discovery of quanta, mathematically curious readers will get a sense of how mathematicians develop their concepts.

Dragan Radulović is a theoretical mathematician who also publishes in the area of molecular biology. He is an ardent traveler and adventurist. His previous book, *On the Road Again - 2018,* is about his road trip through Iran and Afghanistan. After stints at Princeton University and Yale University, he moved to South Florida, where he surfs, writes, and does mathematics.

Why Does Math Work ... If It's Not Real?

Episodes in Unreasonable Effectiveness

DRAGAN RADULOVIĆ

CAMBRIDGE
UNIVERSITY PRESS

Shaftesbury Road, Cambridge CB2 8EA, United Kingdom

One Liberty Plaza, 20th Floor, New York, NY 10006, USA

477 Williamstown Road, Port Melbourne, VIC 3207, Australia

314–321, 3rd Floor, Plot 3, Splendor Forum, Jasola District Centre,
New Delhi – 110025, India

103 Penang Road, #05–06/07, Visioncrest Commercial, Singapore 238467

Cambridge University Press is part of Cambridge University Press & Assessment,
a department of the University of Cambridge.

We share the University's mission to contribute to society through the pursuit of education, learning and research at the highest international levels of excellence.

www.cambridge.org
Information on this title: www.cambridge.org/9781316511923

DOI: 10.1017/9781009051774

First published 2023

A catalogue record for this publication is available from the British Library.

A Cataloging-in-Publication data record for this book is available from the Library of Congress.

ISBN 978-1-316-51192-3 Hardback
ISBN 978-1-009-05481-2 Paperback

To Petrana, my firstborn and my first editor

Contents

Preface

Many years ago, at my US citizenship interview, the very last step was a meeting with a Federal Bureau of Investigation (FBI) agent. Soon after he realized that I was a math professor, he dropped the "agent" part and became a student, and a curious one at that. Very sincerely he told me:

> In college I had to take calculus. I loved it! But nobody explained the big picture to me. How is it possible that the derivative of Sin(x) ends up being exactly Cos(x)? What are the odds of that happening? The area under the curve is the opposite of the slope of a tangent line. How could that be? Why is nobody flabbergasted about this? Did I miss something?

Needless to say, I was very impressed, and we had a great chat – never mind the long line of applicants waiting in the hallway. I also told him that I was writing a book about the very questions he was pondering.

In the pages that follow I address the issue that the curious FBI agent raised. *Yes, we all know that mathematics works and that our universe is guided by the laws of mathematics, but why?* Not only is this an important question, but a surprisingly large portion of people (the fellow FBI agent excluded) do not even realize that there is a question.

You see, very few people know that we mathematicians, by and large, never actually intend for mathematics to be useful, or applicable. Far from it. For the most part, we just doodle with our equations and play with our theorems, ignorant of any applications. And yet, somehow, math is not only applicable, but it seems to govern the very laws of our universe. A miracle if there ever was one.

Why is that? Well, this book will not provide a definite answer. No book can do that. Instead, here, we will work on properly posing this question. We will see examples, many examples, of mathematical formulas that were designed centuries ago, somehow, almost magically, appearing in today's applications. For better or worse, much of the math used by modern science and industry existed for centuries, even millennia.

It almost seems as if some magic hand guided the ancient mathematicians, so that their formulas would be perfectly fitted for the sophisticated technology of today. I know this does not make much sense, and yet this is how it is. This book will convince you of that. It will describe concrete and specific instances in which we mathematicians, completely oblivious to the world around us, created theories that perfectly explain the very world around us. Strange, but true.

The book tackles some serious issues, but it is also designed to amuse and engage the reader. It is written as a series of loosely connected entertaining essays, carefully chosen to convey rather intricate and deep messages. Technical aspects are kept at an approachable level so that the book should appeal to both professionals as well as math enthusiasts. Although some rudimentary knowledge of calculus is helpful, it is not necessary.

For example, I have presented many of these essays in lecture form at local high schools, where students were very receptive and very impressed. These lectures helped me realize that it is possible to address some complex mathematical ideas without overwhelming the audience with complicated formulas and concepts. Consequently, all but the last few pages of this book are accessible to a wide audience: curious high-school students, doctoral candidates, or FBI agents who took calculus a "long time ago."

At the end of the book, in the Post Scriptum and Appendix, I dig a bit deeper and offer more precise mathematics. Here the reader can find the proofs that have been skipped in the main body of the text, as well as some ideas for possible original research. These pages are more challenging and contain some "serious" mathematics, but one can still get the "big picture" here, even if some formulas and concepts have to be skipped.

Finally, I must admit that, as much as I like physics and history, neither of these fields is my specialty. I am trained as a mathematician, and as such, I can vouch for the math presented here. As for the physics and history, well I ask the reader for some forgiveness and some leeway. To the best of my knowledge everything presented here is correct, but a pedantic historian or physicist will likely uncover a strange phrase here or an awkward account there. And for that I apologize.

Based on the advice of my editors and several reviewers I have included some additional resources, available at www.cambridge.org/9781316511923. This material follows the narrative of the book, and it includes numerous short lectures and extended problems. It is designed to prompt critical thinking and to be supplemental material for teachers who are eager to go beyond the standard textbooks.

Acknowledgments

It has always been my dream to write a book – but not this one. English is not my mother tongue, so before attempting a "real" book, I figured I should practice a bit. I picked a topic familiar to me, one that I have some experience with, began writing, and kept on writing: about math, mathematical anecdotes, and interesting bits and pieces of stories I used in my lectures. One thing led to another, and before I knew it, I had a 200-page monster on my desk. That disorganized pile of papers contained the essence of this book, but not much of the form or design. It was never intended for publication. It was created for practice only. To this end, to check my writing skills, I sent it to some of my English-speaking colleagues.

The first one I "terrorized" with the aforementioned "monster" was Eric Foss, a molecular biologist and my collaborator. He loved it, and he asked if it would be OK for him to lend a copy to Franklin Stahl, his de facto stepfather. Frank is the coauthor of the famous Meselson–Stahl experiment, so I was both flattered and frightened. But Frank liked it too, and he said something that changed everything: *I am looking forward to seeing it in bookstores.* Bookstores? It had never occurred to me that what I had written could be published. So in many ways the very existence of this book is Eric Foss and Franklin Stahl's doing – and for this I thank them.

It snowballed from there. I rewrote the "pile" in order to simplify and condense it, and then I tested the newest version on unsuspecting friends and colleagues. Their responses and comments were priceless. Goran Peškir of the University of Manchester had played an important role, so much so that he appears in this book on a few occasions. Ron Hiller, the founder of Quantiva, an internet startup company that I worked for, was another "influencer." So was Slaven Stričević, a mathematician at Pratt & Whitney. Jeffery Fogul, a retired Central Intelligence Agency (CIA) analyst from "my" coffee shop offered some valuable insights as well. I thank them all.

Special thanks go to Željko Ćular-Ćupe, an ocean freighter captain, childhood friend, and *compadre*. He is one of the few who was patient enough to read several versions of this book, and the impact of his encouragement cannot be overstated. (Moreover, a few years back, he wrote a technical book on cargo ship loading, so I figured, if Ćupe can write a book, so can I.)

Stacy Jacobowitz of West Boca Raton High School deserves special credit as well, for she was the first teacher who invited me to talk to her class about interesting mathematics. My interactions with her students did wonders for the narrative of this book. It fine-tuned my essays, making them mathematically relevant, yet not too hard to follow.

Publishing this book was not a bed of roses. I tried myself but nobody would take it. So, back in the drawer it went, for many years. The breakthrough came from an unlikely source: an editor's rejection. Peter Tallack, of the Science Factory, rejected my proposal, but it was a nice rejection. He liked the book, but it was a bit too technical for his taste. He suggested an academic press instead, the Cambridge University Press in particular. And for this I thank him.

I must say that the only reason you are reading this book is because Diana Gillooly, senior editor at Cambridge, liked my proposal. There is no way around it. I was just about to put this "foolish" project aside and go back to "real" mathematics. Luckily, Diana liked it. And she fought for it. And for that I am immensely thankful.

The following people helped with proofreading: Robert Lubarsky (logic), Jaksa Cvitanić (analysis), Vladimir Bozović (algebra), Ata Sarajedini (physics), and Goran Peškir (everything). I am grateful for their help.

Finally, I would like to thank the whole Radulović clan: Serene, Petrana, Tatiana, Tesla, Radovan, and Ratka, for having faith in me and enough patience to put up with me all these years.

PART I

Rare Axioms

1 Introducing the Mystery

1.1 The Magic

It was the summer of 1684 when the magic happened. Or at least that is how I choose to tell this story. Isaac Newton had already developed his theory of gravity and the idea of Copernicus' solar system was now widely accepted. These were some extraordinary achievements, for they had concluded the millennia-long struggle to understand the stars and the behavior of heavenly bodies – a question that we humans have marveled about since the dawn of time. But that was not the magic I speak of.

That fateful summer Newton accomplished another task, rather negligible in comparison: He calculated the *trajectories* of the heavenly bodies. Apparently, Halley had asked him to compute the path of a "certain comet," which Isaac did. Without any measurements or experiments, relying solely on mathematics, he derived the *ellipse*. Later, his calculus brought to light the other possible trajectories, the *parabola* and the *hyperbola*. And that was magical.[1]

For, you see, these three curves, these rather particular types of curves, had been discovered by the Greeks some 2,000 years earlier, and the miracle, of course, is: *How?* Did the Greeks know about the theory of gravity? Of course not. Did they somehow stumble upon these curves by studying the heavenly bodies? Of course not. So, in what context did they uncover these curves, the very ones that even the stars must follow? Well, it's a funny thing actually. The Greeks studied these curves without making any connections to heavenly bodies. In fact, they were for the most part oblivious to any applications of these curves. The curves were discovered as the intersection of a plane and the surface of a cone.

[1] The ellipse was first proposed by Johannes Kepler almost a century earlier, but unlike Newton, Kepler did not have the necessary mathematical apparatus nor the laws of motion to formally, mathematically, prove his experimental observation.

Why anybody would study such a bizarre thing (the cross-section of a cone.) is a different matter but one that leads us to another, more far-reaching question:

> Why would an unrelated mathematical result, conceived thousands of years earlier, play such an essential role in such a fundamental scientific theory? Coincidence? Well, let us see.

The number "*i*" was first contemplated by ancient mathematicians (Greeks and possibly Arabs), and later by Renaissance mathematicians in order to solve quadratic and cubic equations. The number itself is "imaginary"; it does not exist. The very name is derogatory, coined by Descartes, as a way to emphasize its uselessness. Namely, it is "supposed to be," a square root of −1, and as every middle-school child knows *one cannot take the square root of a negative number*. Consequently, for the better part of a millennium it was just another bizarre mathematical construction – until it appeared as an instrumental tool in engineering, both civil and electrical.

Computer architecture is based on "logic circuits," which are just electrical circuits. So why are they called logical? Where is the logic? Well, it is there, hidden within the design. These circuits follow the rules introduced by George Boole, a founder of mathematical logic. Needless to say, Boole introduced his algebra completely oblivious to any technical applications. Computers appeared more than a century after his time.

And there are many, many more examples. It seems that almost all major scientific concepts or theories rely on some mathematical ideas – but ideas developed much earlier and for *different purposes*. This probably comes as a surprise to most, since the majority of people are under the impression that there exists a pleasant synergy between mathematicians and scientists: *Mathematicians* provide the necessary formulas and the scientists, in return, provide the motivation. But this is not the case.

The rigorous scientific method was introduced quite late, during Galileo's time; thus, for most of history, the science did not even exist. And it is very easy to prove this point: Can you think of any genuine law of nature discovered prior to the sixteenth century? Consequently, for a good 2 000 years, mathematicians have been developing their theories completely oblivious to any law of nature. Not necessarily because they did not care, but because nobody knew of any such a law.[2]

Engineers, however have been around for a long time. And they used mathematics. But even with them, we do not see too much interaction. While Thales, Pythagoras, Euclid, and Diophantus were trying to trisect the angle and derive the properties of integer equations, the ancient engineers went on building their structures, and there is very little evidence that they ever utilized these results. Similarly, during the Middle Ages, al-Karajī worked on the binomial theorem, Fibonacci was playing with his numbers, while Cardano and Tartaglia were developing algebraic solutions for third-degree polynomials. But, yet again, the great engineers of that period were not aware of these results.

[2] There were many laws of nature proposed during ancient times, but it seems that only the laws of Archimedes stood the test of time.

Of course, the Arab engineers, and later the Renaissance engineers, used mathematics, but not contemporary mathematics. By and large, they both used Greek mathematics, developed 1,000 years earlier. It continued like this for a while, until that fateful summer of 1684, when Newton, with the stroke of a pen, bridged the 2,000-year gap. This synthesis did not go unnoticed. For around 100 years, there was a great conciliatory period, resulting in an avalanche of mathematical results inspired by physics (and vice versa): differential equations, dynamical systems, vector spaces, harmonic analysis ... and the list goes on.

But alas, it did not last. It did not take long before mathematicians went astray and, in the Greek tradition, started working on "irrelevant" things again - weird things like how to define an infinite dimensional space, or how to construct functions that are discontinuous at every single point, or what is the volume of a unit ball in five dimensions. Some went back and revisited the very foundations and redefined obvious things like addition and multiplication. Even counting was deemed overdue for revision. In their epic work of the early twentieth century, Bertrand Russell and Alfred North Whitehead proved that $1 + 1 = 2$. It took more than 100 pages.

Needless to say, scientists did not find these results compelling, so we parted again. By the end of the nineteenth century, there was very little overlap between math and physics. In fact, following the death of Gauss, the percentage of mathematicians producing significant results in physics (or vice versa) diminished very quickly. So, things did indeed revert to the good old Greek days.

But the "coincidences" did not abate. No matter how peculiar and unusual the objects mathematicians studied, somehow, whether a few decades or a few centuries later, scientists would come up with a requirement that seemed tailor-made for this weird math. And strangely enough, more often than not, it was the "bizarre" math that turned out to be more applicable in the long run, much more than the "applied" math designed specifically to help engineers and scientists.

Here is another example. Mathematicians did not quite like Euclid's Fifth Axiom of Geometry. There was no rational explanation for this distaste. Somehow, the axiom was felt to be clumsy, too complicated. For this reason (and only for this reason) people tried to eliminate it or to replace it with something more elegant. They could not remove it, but strangely enough they could replace it. Lobachevsky and Bolyai did so and, as a result, developed some wacky geometry in which ridiculous things could happen: two parallel lines would eventually intercept, angles inside triangles did not add up to 180°, and so forth. This geometry was completely useless for it did not relate to anything real, and one could only wonder why any sane person would spend a lifetime studying such an odd thing. Well, it turned out (some 100 years later) that Einstein's theory of relativity implied that space-time is curved and, consequently, has many of the "bizarre" geometric properties predicted by Lobachevsky and Bolyai's weird construction. *Now this is one big coincidence!*

This mystery did not go unnoticed. How could it? Einstein himself wondered: "How can it be that mathematics, being after all a product of human thought independent of experience, is so admirably adapted to the objects of reality?" Leonardo and Galileo and Dirac, and many others, had noticed it too. They offered some clever maxims, but they did not follow up with a thesis or explanation. Wigner's essay on the *unreasonable effectiveness* of mathematics is probably the best-known attempt to explain the phenomenon, but his treatise does not go into great detail, and in many ways its goal was to explain away the great mystery. Among philosophers, historians of mathematics, cognitive linguists, psychologists, and such, one can find a sizable body of work related to this topic. A large number of professional thinkers have tackled this problem.[3] As they should have.

Nevertheless, among the vast army of people exposed to mathematics - people who create mathematics, people who apply mathematics, people who teach mathematics - this topic remains unacknowledged. Many of us, from an early age, couldn't help but notice this mysterious action of mathematics. We all saw π and e and *Sin* popping up in the most unexpected places. But our teachers did not stop to explain. Textbooks did not stop to explain. Nobody did. Instead, they just plowed forward through the curriculum, the ever so convoluted curriculum. And mathematics, like any other art form, requires attention and proper display.

If one climbs higher and learns more of mathematics, things do not really change. Far from it. The bizarre mathematical connections multiply, even explode, but explanations are hard to find. At the very top of the pyramid, among professional mathematicians, this mystery is rarely mentioned, and almost never discussed. In many ways it is a hush-hush topic. We are obviously aware of it, but then again, who has the time? We have other things to worry about. We have to apply these magical formulas, or we have to discover more of them. Seldom, if ever, do we ask: Why are they here? We leave these questions for philosophers to ponder. For, who in his right mind would risk his research career and write a book about it?

Well, here it is. The book. Which is not necessarily about the answers but more about the question. And this question, I believe, has never been properly presented to a wide audience. I will explore, in some detail, this mysterious behavior of mathematics. I will demonstrate how it imposes itself on us, and how it does so in the most surprising ways. You study the stars and mathematics is there, you look at DNA and mathematics is there, you randomly toss a coin and mathematics is there, you examine a network of friends on the Internet and mathematics is there. And all too often its appearance is not linear or expected, but completely unanticipated.

I will show examples where mathematicians created formulas and concepts specifically designed *not to be useful*, and yet, a century later, even these formulas would somehow, magically, embed themselves into the very fabric of our lives. Thus, in many ways, this book will deepen the mystery. At the end, I am afraid, it is likely that the reader will feel as confused as ever, but I believe confused on a higher level and about more important things.

As for the answers ... I offer a few, but I must admit that it is easier to ask a question than to answer it. Nevertheless, asking the right question is an important task on its own and I hope the reader will appreciate it.

[3] Lakoff and Nunez attempted to decipher the cognitive origins of math. The mysterious connection between physics and math has been well explored in the collection of essays/papers "Trick or Truth." Philosophical issues related to math applicability are covered by Steiner. Even Badiou writes about math, set theory, and the eternal truths. Among the more popular books, this topic has been addressed by Hamming, Tegmark, and Farmelo.

Disclaimer

In this book I will often say "we mathematicians" or "we" or "us," which deserves an explanation. As mathematics is arguably one of the oldest continually practiced intellectual endeavors,[4] and as it has evolved into an incredibly complex and gargantuan entity, it would be absurd of me to speak for all mathematicians. That said, there exists a significant subset of mathematicians for whom I do speak. Many of my sentiments and opinions regarding the role of mathematics and its "usefulness" are strongly influenced by G. H. Hardy's book *A Mathematicians Apology* (Cambridge University Press, 1967, p. 119). Take the following few lines:

> The "real" mathematics of "real" mathematicians, the mathematics of Fermat, and Euler and Gauss, and Abel and Riemann, is almost wholly "useless" (and this is as true for "applied" as for "pure" mathematics). It is not possible to justify a life of any genuine professional mathematician on the ground of "utility" of his work.

Many professional mathematicians will disagree with this statement. But then again, many will not. And in order to avoid unnecessary repetition, from now on, when I say "we mathematicians" I will mean mathematicians who agree with Hardy's statement. As for those who do not, well, they are still "real" mathematicians of course, but just hold a different opinion.

1.2 Obvious, Irrelevant, and Just Right Math

A joke: Two friends get lost while flying in a hot air balloon. After several hours of aimless drifting in fog and at low altitude, suddenly in a clearing they see a hiker on top of a hill. "Where are we?" shouts one of them. The hiker on the hill looks at them, pauses for a second and answers: "In a hot air balloon." Seconds later a wind blows them back into the fog.

> "That person is a mathematician!" the first balloonist says. "Why a mathematician?" the second asks.
>
> "Well, before responding the hiker thought very carefully, the answer we got is an absolute truth, and ... it is completely useless."

My childhood hero, the famous physicist Richard Feynman, allegedly said: "Mathematicians spend lots of time proving things that are either obvious or irrelevant." I could not agree more. Mathematical theorems can be separated into three

[4] Math problems referencing quadratic equations and a version of Pythagoras's Theorem appeared in Mesopotamian documents, about a thousand years before any writings of Socrates, Buddha, or Confucius. These math problems are near identical to the ones we find in today's textbooks.

groups: *the obvious, the irrelevant,* and *the just right* (useful but not trivial). The funny thing is that most of our results belong to the two former, "useless" categories.

An example will help.

Let us start with basics, the bare basics. Suppose you were to build a 10-foot-high roof for a house that is 20 feet wide. To finish the roof, you need the rafter, that is, the diagonal beam on which the tiles will be laid. You need to cut the beam to some length, not too short and – since this rafter will be rather heavy – you also need to be sure that it is not too long. So, how long should it be?

One way to tackle this problem is to scale it down and *make* a model drawing of the house. If we scale it down by factor of 10, we get a model whose roof is 1 foot high and 2 feet wide. All we need now is to carefully measure the diagonal (which is approximately 1.41 feet) and then scale it back up tenfold. The result yields a 14.1-foot-long beam. Perfect!

The obvious drawback here is the "scaling" part that needs to be repeated for different roofs. We need a more general tool, and it is provided by a formula: $a^2 + b^2 = c^2$, which in our case yields $c = \sqrt{1+1} = \sqrt{2} \approx 1.4142$. With this formula we do not need scaling. It works for *any* roof, however wide, however high.

A trained mathematician will quickly detect the theorems at work here, but I challenge the reader to recognize the parts where mathematics plays a role and how the applications intermingle with it.

There are actually two theorems here. First is the theorem on similar triangles, guaranteeing that the scaling works as we want, and second is the theorem of Pythagoras, giving us the formula. The first result states that *two triangles have sides that scale by a fixed factor if and only if their corresponding angles are identical.* This seemingly trivial fact yields some rather surprising mathematical applications. It is how we know the height of Mount Everest, for example.

The second piece of math in our roofing story, the celebrated Pythagorean formula, is a well-known result and, for this reason, often taken for granted. But I urge you to think about it for a moment. What right do we have to expect such a general result? For several millennia (literally between 4000 BC and 1000 BC) ancient builders constructed magnificent monuments, pyramids, aqueducts, and mausoleums. In the process, they developed a myriad of tricks and rules of thumb. They produced scaled-down drawings for various roofs and triangle constructions. They came very close to discovering this equation, for they had tables of Pythagorean numbers *a, b,* and *c* that satisfy the above formula, but they did not know the equation.[5]

They were certainly very smart and resourceful people, and I am confident that they would have been surprised to find out that all those sketches were superfluous. The Pythagorean formula applies to any roof (any right-angled triangle), no matter how strange-looking. It is astonishing that such a unifying formula exists at all! Why would it?

[5] There is near consensus among historians that the Egyptians did not know the full power of the Pythagorean Theorem. As for Mesopotamians, the jury is still out. The actual proof, or references to the proof, is still missing, but some indirect evidence suggests that the formula was known (or guessed).

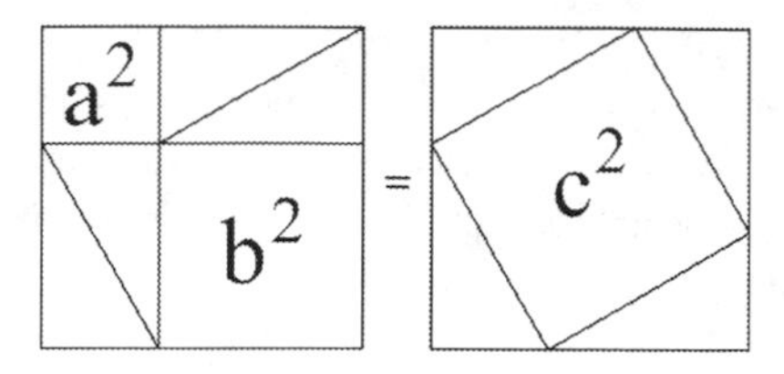

Figure 1.1 A proof that $a^2 + b^2 = c^2$.

Surprisingly, not only does such a formula exist, but the proof is very elegant, and it is worthwhile including it (in Figure 1.1). Once you "see" it you cannot believe how simple it is, and yet, for thousands of years engineers were not aware of it.

Both squares have area $(a+b)^2$ and both squares contain four identical right-angle triangles - with sides **a** and **b** and hypotenuse **c**. Therefore, after removing the triangles, what remains is the Pythagorean formula $a^2 + b^2 = c^2$.

Before we wrap up our roofing tale, there is one more story to tell. It has to do with $\sqrt{2}$. This number appears as a natural consequence of the Pythagorean formula, since for a unit square and its diagonal *c* we have

$$c^2 = a^2 + b^2 \Rightarrow c = \sqrt{1+1} = \sqrt{2},$$

and the ancient Greeks had to accept it, but they did not like it. They believed that any number should be expressible as a fraction, that is, as a ratio of two integers, like ½ or ¾. This is a reasonable assumption, for it seems that any decimal number can be trivially written as a fraction. Say you have the number 3.44. It equals 344/100 = 86/25.

The mathematical methods of the ancient Greeks relied heavily on proportions (e.g., the scaling trick), and they had developed a kind of fixation here (the golden ratio). Consequently, they convinced themselves that it is rational to assume that any number can be expressed as a ratio of two integers. Notice that the words *ratio* and *rational* have the same root, which is surprising since their meanings, *fraction (or rate)* and *reasonable*, are not related, except that - as it happens - the ancient Greeks believed that fractions are reasonable.

But the Greek mathematicians were fair. Unlike other religions,[6] which encourage blind faith, math demands a proof. In this case, a proof that every number can be expressed as a fraction. But this proof cannot be found. Instead, a counterexample was produced. They (the authorship has been lost) proved that the aforementioned $\sqrt{2}$ could not be expressed as a fraction (although 99/70 = 1.414285 comes very close). Numbers that cannot be expressed as a fraction are called irrational (the "unreasonable" numbers). The proof of this astonishing theorem is very easy, all one needs to show is that if $\sqrt{2} = n/m$, then both n and m must be even, which is a contradiction (see the Appendix for details).

[6] Yes, mathematics has all the attributes of a religion. We follow it diligently, and we all believe that it is consistent, without contradictions. And, as with other religions, we cannot empirically justify this belief.

Now we can address the title of this chapter. The roofing tale above yielded three theorems: similar triangles (obvious), Pythagorean formula (just right), and the irrationality of $\sqrt{2}$ (irrelevant). I think Feynman would approve. The last theorem is irrelevant from a physicist's point of view. It has been more than 2,500 years now, and nobody has found any practical application related to the irrationality of $\sqrt{2}$.

Feynman was right in other respects, too. The vast majority of mathematical results fall into the two "useless" categories, and the reader should not take my word for it. A simple glance through math journals, even very old ones, will reveal few applicable results. Or consider the list of all Fields Medalists (the equivalent of the Nobel Prize for mathematicians) and you will find very little, if any, applied mathematics among their body of work.

Unless the reader is a mathematician by profession, she will have a hard time accepting this fact. A plausible explanation comes to mind: Mathematicians are most likely looking for useful theorems, but since such results are rare, inevitably they produce a lot of "useless" theorems along the way.

Not true. The majority of us are not looking for applicable theorems. We are trained to seek universal truths, deep mathematical results that will help us understand mathematical theory. If some of our theorems turn out to be useful elsewhere, so much the better, but that is not our primary goal.

And no, this puritanism is not an invention of the twentieth century. The founders of mathematics, the Greeks,[7] were obsessed with pure reason. They held logic and reason superior to our "feeble" senses. Measurements and experiments inevitably depend on our senses and were deemed inferior to reason. Measure the triangle as much as you want, but you will never be absolutely certain that the formula $a^2 + b^2 = c^2$ is true. (The author tried this and could never get within 1% error.)

The three famous problems of antiquity (*squaring the circle, trisecting the angle,* and *doubling the cube*) played central roles in Greek mathematics, but each of these problems is unsolvable and thus, from a practical point of view, useless. But this has not stopped scores of mathematicians from working on them. In the process they developed a fantastic number of beautiful theorems and laid the foundation for mathematics to come.

Or how about the story of amicable numbers? This is one of the pearls of useless math. Two numbers, N and M, are called amicable (friendly) if the divisors of N add up to M and vice versa. An example will help.

Let $N = 220$. Its divisors are

$$1,\ 2,\ 4,\ 5,\ 10,\ 11,\ 20,\ 22,\ 44,\ 55,\ 110,$$

and if we add them up, we get

$$1+2+4+5+10+11+20+22+44+55+110 = 284.$$

[7] Here I refer to mathematics based on proofs and axioms. The available records strongly indicate that it was the Greeks who first introduced and develop this form of art.

Now try the same with 284. Its divisors are

$$1,\ 2,\ 4,\ 71,\ 142,$$

and they add up to

$$1+2+4+71+142=220.$$

Nice. Therefore, 220 and 284 are amicable numbers. How about finding a few more? One quickly realizes that this is not an easy task. The ancient Greeks knew about this pair (220 and 284) but could not produce any other. It took another thousand years or so before Arab mathematicians produced another pair (17,296 and 18,416).[8] Europeans tried too, Fermat and Descartes in particular. When all is said and done, by the eighteenth century, mathematicians had managed to produce just a few amicable pairs. Two thousand years of work and only a handful of pairs produced. And not for lack of trying.

Needless to say, there are no applications involving amicable numbers, and there are very few mathematically relevant results related to these numbers. So, if this does not strike you as a colossal waste of time, then nothing will. The story does not end with Descartes, of course. Euler managed to produce dozens more, and amazingly enough, in 1866, a 16-year-old boy (with a famous name), Nicolò Paganini, produced another pair of very simple amicable numbers that had eluded the mathematical heavyweights for millennia. His numbers are 1,184 and 1,210.

If you are still not convinced, if you still believe that mathematicians aim to help engineers and that engineering played a fundamental role and influenced the Greek mathematicians, consider the following: The Romans were great engineers, but the Greeks were not. The Romans built aqueducts, highways, bridges, but it is hard to find one, just one, well-known Roman mathematician.

Greek engineers had a fantastic advantage. Had they actually studied and applied the wonderful mathematics that their contemporaries provided, they would have been able to engineer amazing structures that the Romans and Egyptians could have only dreamed of. However, there is no evidence that they did that.

The Greeks knew the properties of parabolas and could have produced very powerful reflectors. Their mastery of geometry was superior to that of Romans, but we do not see that reflected in the structures they left behind. Although the Greeks produced some inspiring buildings, these pale in comparison to those built by the heavyweights: the Romans, the Egyptians, and the Chinese. There are no Great Walls of Greece, no Great Pyramids of Greece, no Greek Colosseum, highways, or aqueducts. They were philosophers, not engineers. It was Appius Claudius who allegedly said: "Alexander conquered the whole world; imagine what he could have done had the Greeks known how to construct a Roman road."

[8] There is ample indirect and anecdotal evidence showing that Arab and Persian mathematicians produced a few more of these numbers. However, well-documented sources are missing.

1.3 Why Did They?

After these few pages, I hope the reader is intrigued. Apparently, mathematics, both modern and antique, is not designed to help scientists and engineers. And we mathematicians, both contemporary and of the past, study mathematics for mathematics' sake, because we find it beautiful, intriguing, and interesting, which makes sense once you learn how to fully appreciate the art. To quote Hardy: "Beauty is the first test: there is no permanent place for ugly mathematics." However, a natural question arises: Why did the Greeks – the champion philosophers, the founders of this art – create mathematics? Surely not to help engineers, for it is difficult to see why ancient engineers would care about the cross sections of a plane and a cone.

To answer this question, we have to go back in time almost three millennia and imagine a young scholar (let us call him Petros), studying the great philosophers of his time. He reads Aristotle and finds his assumptions appealing, his logic impeccable, and his conclusions interesting. But Plato's assumptions and reasoning, as well as his conclusions, are interesting too, and so are Zeno's and Socrates'. But there is a problem of course. The conclusions of these great men are often in disagreement.

This puzzles our young scholar. How can this be? How can these great men come to contradictory conclusions? The problem is not with their logic or reasoning, for the great philosophers were impeccable there. Something must be wrong with their assumptions. No matter how plausible, they cannot all be correct.

So Petros and his followers decide to do things differently. They try the following psychological experiment. They start with assumptions so trivial that nobody could ever dispute their correctness. Statements like $2 + 2 = 4$ or *a triangle has three angles.* Surely nobody could question these. Then, with these assumptions as a starting point, using only logic and reason, they try deriving as many conclusions as possible. This process should inevitably produce the absolute truths that all the great philosophers would agree on. The only problem, reasons Petros, is that these "absolute truths" were likely to be trivialities as well. If one starts with trivialities, one cannot get too far, right?

But, nevertheless, they tried. Using trivial assumptions like "a circle is defined by its center and its radius" and "through two points one can draw only one line," they began deriving as many logical consequences as they could. These "absolute truths" were called theorems. Many of these theorems were still trivial statements (this is precisely the source of "obvious" mathematics), but surprisingly, some were not. And the more they tried, the more these surprising and unexpected results kept showing up.

For example, take a triangle and a circle. There is nothing in their definitions or appearance that points to any connection between them. One is "spiky" and "crooked" while the other is "perfect" and "round." However, the Greeks have shown that these two objects are intimately connected. Take three points (not four or five, but three) and place them anywhere, and miraculously, there exists one (and only one) circle that can be drawn intersecting these three points – as long as these three points form a triangle. This result is nothing short of astonishing, and if the reader thinks it is trivial, I challenge her to prove it.

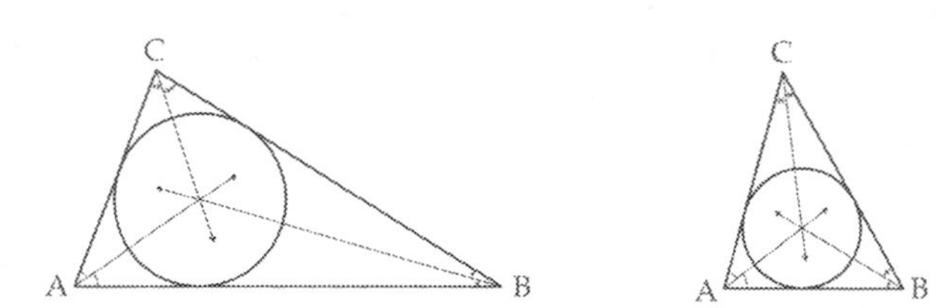

Figure 1.2 Triangles and their inscribed circles.

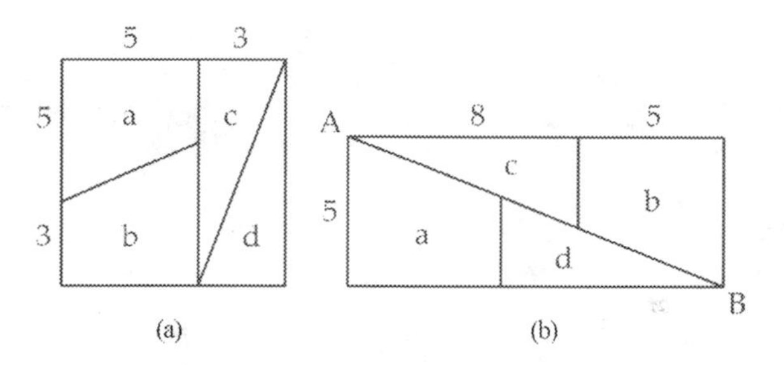

Figure 1.3 "64 = 65. Our feeble senses play a trick on us"

Another example: For any *triangle,* no matter how irregular, its three angles always add up to half a *circle.* And another example: Take a triangle and bisect each of its three angles. The three lines you get will always intercept in a single point. And this triangle can be of any shape or size. As if this is not enough, this point of interception has special properties: It is the triangle's center of mass and it is also the center of the inscribed circle. Figure 1.2 shows two examples. The reader might be tempted to pooh-pooh these results as something children learn in middle school, but the Romans and the Egyptians, the master engineers of antiquity, did not know them. The reader is more than welcome to try to prove these "simple" results.

In mathematics, *to see* is not the same as *to believe.* Consider the little puzzle laid out in Figure 1.3. The four polygons *a, b, c,* and *d,* if rearranged into a square (Figure 1.3a), produce an area of 64, while if rearranged into a rectangle (Figure 1.3b), produce an area of 65. Clearly $64 \neq 65$, so something is amiss. Our feeble senses have played a trick on us. There must be some mistake, and it is not easy to spot. This particular example illustrates why mathematicians proceed with extreme caution. In our world, statements like *beyond a reasonable doubt* do not carry any weight. Every argument must be perfectly correct, proved *beyond any doubt whatsoever.*

The Greeks constructed an intricate maze with thousands of theorems, all in perfect harmony, none ever contradicting the other. Most of these results were too sophisticated for ancient technology to use, and they were left idle, even forgotten.

But then something magical happened. As centuries passed, the majority of these "useless" results were rediscovered (with substantial help from the Arabs and the Persians) and implemented in modern science and technology. Often, they even played an essential role.

And this is the central point. This is the very question I pose and will try to answer. *Why does mathematics work?* The story of Greek mathematics is not the only one. We see these examples over and over through history. Some seemingly unrelated mathematical theory plays a central role in scientific discovery centuries, even millennia later. How is this possible?

Disclaimer

I took some liberties with Petros's story. We know from historical records that axiomatization – Petros's strategy of starting with indisputable assumptions – first proposed by Euclid, occurred centuries after many of the theorems were proven. So, in a way, the real story proceeded in the opposite chronological order from what I described. Nevertheless, I believe that the early mathematicians, in particular Thales and Pythagoras, acted out Petros's story in spirit if not to the letter. The axioms were most likely deemed so self-evident that mathematicians did not even bother formally stating them. There is no evidence that they started this process the way Petros did, and I admit my arguments are completely speculative. I only hope that the reader will find them intriguing.

2 On Classical Mathematics

Let us revisit some of the aforementioned bizarre mathematical coincidences; the magic if you will. Maybe, if we take a more careful look, if we dig deeper, we will uncover the culprit, the mechanism that binds mathematics and science together.

2.1 The Stars and the Curves

2.1.1 The Stars

When one thinks of a farm one might paint an idyllic picture and imagine the geese, the sheep, the corn, and the turkeys, kids playing in a backyard, and an apple pie on the kitchen windowsill. Or one might be more realistic and think of the harsh manual labor, the mud and the dust, and the pre-dawn working hours. Either way, one does not typically associate farming with sophistications, and with civilization. But the facts are irrefutable: We owe our civilization, its very existence, to farmers.

It is very simple actually. Once the prehistoric nomads figured out how to plant cereal grain, the necessity for hunting and gathering declined. With bountiful harvests the population increased and nomadic life was replaced with a more stationary, farm-oriented existence. As harvests became more reliable, the first permanent settlements formed, and then the first cities, and inevitably civilization.

But these first communities were fragile. The early crops were imperfect and the harvests varied in size, resulting in frequent famines and the demise of many cities. Several factors played a role: the seeds, the technology, and the climate – all things that the early humans could not do much about. However, one important factor was under human control, the calendar.

It is probably difficult for modern readers to comprehend the importance of a good calendar. But think about the fact that the very survival of one's tribe, the lives of all your friends and relatives, depends on the harvest. Imagine now that you do not know that a year has 365 days. It is not hard to foresee the consequences. Sooner or later, you will make a mistake - a big mistake. A few unseasonably warm weeks in late winter would prompt you to plant the seeds too early, assuming that the spring has come, which in turn could result in a catastrophically poor harvest and consequently a famine. And a famine like that could easily wipe out the whole tribe. And very often it did.

Help came from the heavens. For it is there, within the stars, that a clever man discovered an order, and constructed the first calendar. This was not an easy task since the order was not apparent. The stars follow a very zig-zag path, appearing and reappearing in a seemingly random way (Mars, at times, actually travels backward through the night sky). It took a lot of patience, ingenuity, and hard work by generations of stargazers before an accurate calendar was produced. We know now that essentially every civilization on Earth has produced a calendar (the Egyptians, Mesopotamians, Incas, Mayans, and Chinese) but we do not have written records documenting the early stages.

There is a good reason for this lacuna. In order for civilization to prosper, we needed stable harvests, which in turn meant that we needed a calendar. Thus, the construction of a good calendar predated civilization itself. And that is the reason we do not have many written records about early calendars. They had developed before civilization had.

It would be interesting to know what kind of social obstacles early astronomers faced. They were in direct competition with the prehistoric shamans and medicine men, who insisted on their "primitive" gods and demanded sacrifices. These gods and these rituals could not help the harvest of course, but tradition dies hard and one can only imagine the battles that early astronomers fought - the first clash between "Church" and "Science." But the astronomers won, since they could deliver. They could tell farmers when to plant and when to harvest.

These "wise men" replaced the shamans, and they turned into priests. They became very important and for next few millennia the heavens and the stars remained in their "holy" (Church) hands. For indeed, Mayan calendar makers were the priests, and so too were the Egyptian. The Catholic Church kept a firm grip on the study of the heavenly bodies as well. The very word *heaven* is a synonym for *sky*. For the sky is where the gods live, and it is there where our souls would go.

Unlike the primitive cultures with their earthly gods, the gods of wind and earth and water, the advanced civilizations were focused on the stars and heavenly bodies. The "advanced gods" were always found up in the heavens, among the stars. This fixation is directly related to the calendars and their profound effect on early civilizations. The movements of heavenly bodies have left a deep presence through our culture. We all measure time in weeks (the length of moon phases) and divide the year into months (roughly one lunar cycle). Our "right angle," or "quarter of a circle," is assigned the measurement 90°. (Why 90? Because a full circle is assigned 360°, which was the first estimate of a calendar year.) Our destiny is "written in stars," and we are all looking for our "lucky star."

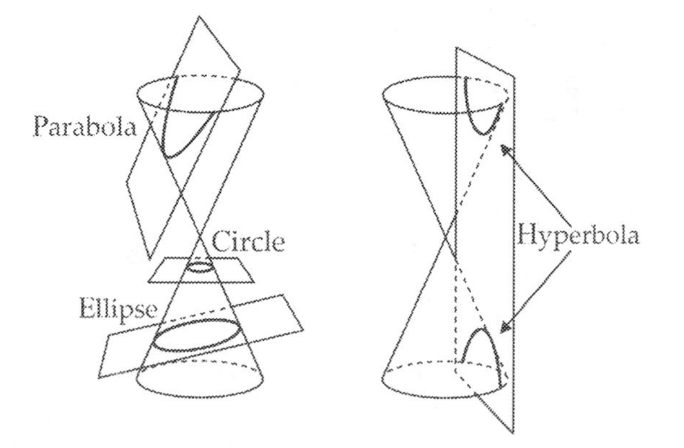

Figure 2.1 Cutting a cone with a plane.

2.1.2 The Curves

Completely oblivious to this stargazing, Greek mathematicians worked on their favorite items: circles and triangles. They did an outstanding job, discovering essentially all the relevant properties of circles and triangles, and leaving only a very few theorems for later generations. The first serious excursion into the "world of curves" is attributed to Menaechmus (4 BC). In his attempt to solve one of the fundamental mathematical problems of his times, *doubling the cube,* he considered the cross sections created by a plane and a cone.

Menaechmus discovered that if one takes a plane and "cuts through a cone," one can get three different curves: the *ellipse,* the *parabola,* and the *hyperbola,* as we see in Figure 2.1.

These are very specific and delicate curves. Take, for example, the ellipse. At first glance it is nothing more than a squished circle, an oval. But not just any oval. If one enlarges an ellipse with a curve that is a unit distance outside the original ellipse as in Figure 2.2, one gets an oval that looks very much like an ellipse, but it is not. Why? Because it cannot be produced as a cross section between a plane and a cone, that is why. Not all ovals are ellipses. Far from it.

It took an additional hundred years or so before Apollonius of Perga finally decrypted most of the mysteries behind these three curves. The connection to the doubling of a cube was a dead end, but he did produce an array of very interesting properties. For example: An ellipse, unlike a circle, is defined by two centers, called the foci. The parabola, on the other hand, is determined by one point (the focus point) and one line. Strange, isn't it? This focal point has the very property that its name suggests: If one makes a mirror in the shape of a parabola, the point where all the sun's rays would converge is exactly that focus point of the parabola. No other curve has such a property.

The ancient Greeks described many other interesting properties: tangent lines, areas, and so forth. What they did not do is apply these results. They actually never even bothered graphing these curves in a careful fashion. A simple and elegant

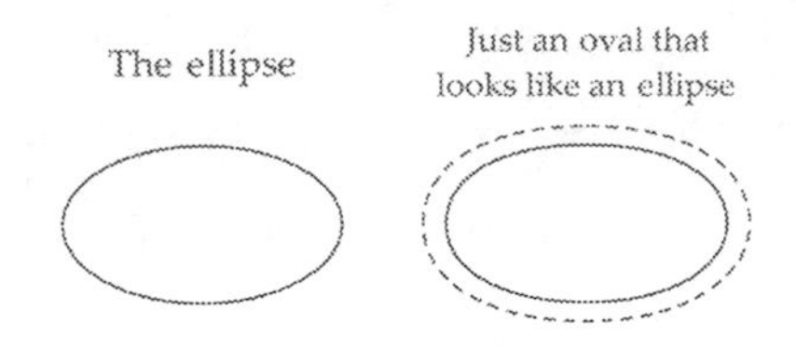

Figure 2.2 Not just any oval.

modification of the usual compass, proposed by the Persian mathematician Abū Sahl al-Qūhī, some thousand years later, is capable of drawing these curves. But the Greeks never discovered it. This was not their cup of tea, for they treasured only logic and pure reason. The actual drawing and construction were left for engineers to worry about.

2.1.3 The Finale

With Copernicus, Kepler, and Newton, the millennia of stargazing and careful mapping came suddenly to a halt. One does not need complicated and imperfect models to determine the exact time and the positions of particular planets. The movement and positions of all the heavenly bodies are easily explained by a model children learn in kindergarten, and all the predictions regarding the positions of the planets can be made by applying Newton's inverse square law. The grand finale, and arguably one of the biggest surprises witnessed by man, comes with Newton's calculations of the earth's trajectory. It is an ellipse, with the sun at one of its foci.

But how could this happen? This very, very, specific curve, discovered 2,000 years earlier in a completely unrelated context, turns out to describe the essence of our planet's journey, its trajectory around the sun. How could the Greeks have just stumbled on these curves when there are literally hundreds of other curves to investigate? What could possibly connect a cone and the earth?

2.2 The Angles and the Music

2.2.1 The Angles

It is well known that the triangle was the epicenter of ancient Greek mathematics. This itself comes as a surprise, for one would expect that some other, much prettier and more elegant geometric shape like a square or pentagon or hexagon would be more mathematically appealing. But they were not. It was triangle that ruled the mathematical kingdom, since it yielded many interesting and important theorems. And one cannot study a tri*angle*, without studying the angle.

Very early on, mathematicians, and astronomers, devised a way to measure angles. The well-known convention of 360° for measuring a full circle was later enriched with a simple, almost trivial, idea of taking the ratios of the sides of a right-angled triangle:

$$\text{Sin}(\alpha) = \text{opposite side/hypotenuse.}$$

$$\text{Cos}(\alpha) = \text{adjacent side/hypotenuse.}$$

The early versions of these *trigonometric* functions appeared in Greek mathematics, but it took another 2,000 years before the Arabs perfected them.[1]

From a mathematical point of view, the main reason, and probably the only reason for the initial development of trigonometric functions was the triangle. For example, consider these two beautiful formulas,

$$c^2 = a^2 + b^2 - 2ab\text{Cos}(\gamma) \text{ and } \frac{\text{Sin}(\alpha)}{a} = \frac{\text{Sin}(\beta)}{b} = \frac{\text{Sin}(\gamma)}{c}.$$

The first, the famous *law of cosines*, is the generalization of Pythagoras's Theorem. It connects a^2, b^2, and c^2, but it is applicable to *any* triangle and not just right-angled ones. If the triangle in question is right-angled, then $\gamma = 90^\circ$, $\text{Cos}(90^\circ) = 0$, and our formula becomes the Pythagorean formula:

$$c^2 = a^2 + b^2 - 2ab\text{Cos}(\gamma) \Rightarrow c^2 = a^2 + b^2 - 2ab\text{Cos}(90^\circ) \Rightarrow c^2 = a^2 + b^2.$$

The second formula above, the *law of sines*, is very useful too. Among other things, it allows us to *triangulate the distance*. With this formula we can deduce the distance to an object that is out of our reach, say an enemy's big guns.

But of course, mathematicians could not just stop with the simple triangle applications. They quickly discovered that trigonometry was fertile ground, an area crammed with hundreds (maybe even thousands) of nice theorems. So, they went on harvesting. They obtained formulas for half angles and double angles and triple angles. They came up with formulas connecting the squares and the cubes of various combinations of various trigonometric functions. They discovered a whole new universe, with a myriad of equations and formulas, some of which were actually useful.

The original definition and motivation for trigonometric functions does not really work for angles greater than 90°, but that did not stop mathematicians. For example, the formula $\text{Sin}(3\alpha) = 3\text{Sin}(\alpha) - 4\text{Sin}^3(\alpha)$ works for $\alpha = 70^\circ$ but it requires a definition (interpretation) for $\text{Sin}(3\alpha) = \text{Sin}(210^\circ)$. There are no triangles with 210° angles, which should create a pickle. In the end, the formulas were too good to ignore and mathematicians decided to preserve the formulas and extend their explorations beyond the triangle.

The first trigonometric extension was made for angles up to 180° (thus covering all the triangles) and soon after to the circle 360° (covering any planar figure). These extensions had to be done very carefully since the original geometric intuition for Sin (as *the ratio of opposite side and hypotenuse*) does not apply anymore. But why stop at 360°? We can easily extend trigonometric functions to any angle by just "cutting and pasting" the 360° function. Thus, $\text{Sin}(x)$ is defined for any number x. All the formulas still apply, but it is safe to say that in the process we have completely lost the original geometric intuition. Angles larger than 360° do not exist in planar geometry,

[1] Many nations played a role in this trigonometric saga: Egyptians and Babylonians, as well as Indians and Persians.

and negative angles plainly make no geometric sense. Intuitive or not, these formulas are interesting and for this reason (and essentially only for this reason) they were studied and studied and studied ....

2.2.2 The Music

Very early on, people noticed that some musical instruments resonate. If one tunes a stringed instrument properly, one can induce vibrations of its strings without ever touching them. By playing a proper note nearby, one can make the strings "pick up the music" and move on their own. This spooky "action at distance" must have been an unusual experience for ancient people, but we know now that it is just one of the many strange properties of waves.

Musicians explored this phenomenon and discovered an orderly and well-defined world. They organized it and introduced musical scales, notes, measures, beats, and the like. Thus, strangely enough, it seems that musicians were the first to discover an order in the natural world, and since they studied it, they should be considered the earliest scientists. Our surroundings are chaotic and random, and it took millennia before we discovered the laws of nature. For ancient people there were only two sources of perfect order: the stars and music. Both behave in a complicated but very predictable and orderly way. The Greeks were fascinated by this harmony (pun intended).

It took another few thousand years before modern science finally explained this strange phenomenon. The strings have their natural frequency that can be tuned using tension. Sound is a wave that travels through the air, and if the sound (i.e., the wave) is of the right frequency, it can excite the strings, by bouncing off them. Physicists soon discovered that there are other waves besides sound waves. In our modern world we have radio waves and microwaves, while in natural settings we have electromagnetic waves and their derivatives: light waves, UV waves, and infrared waves. Cosmic rays, gamma rays, and X-rays are also examples of waves.

2.2.3 The Finale

The reader is probably already guessing what is coming next. Somehow, miraculously, the vibration of a string is related to the ratio of a triangle's sides, the $\mathrm{Sin}(x)$ function. I am sure that many readers are familiar with this fact, but I am also convinced that most readers do not fully appreciate the gravity of this statement - or its consequences. It turns out that for some cosmic reason, all waves, whether a *beach swell*, a *C-note*, or *UV rays*, when their amplitude is plotted against time, produce the $\mathrm{Sin}(x)$ function.

This is an astonishing turn of events since we know that the motivation behind trigonometric functions has nothing to do with waves. For a thousand years mathematicians did not even know how to graph the $\mathrm{Sin}(x)$ function, let alone realize that it resembled a wave. To make things even stranger, the three aforementioned waves (a beach swell, a C-note, and UV rays) are produced by completely different physical processes and propagate in different media: water, air, and a vacuum. Nevertheless, somehow, they all have the same function for their amplitude, the $\mathrm{Sin}(x)$ function. Not *approximately* a $\mathrm{Sin}(x)$ function but *exactly* the $\mathrm{Sin}(x)$ function. A miracle if ever there was one!

2.2.4 *Since We Are on This Topic ...*

Arguably, one of the most bizarre developments in modern science is related to wave-particle duality. With Newton's prism experiments we had confirmation that light is a wave, but with Einstein's explanation of the photoelectric effect we went back to square one, for it seemed that light is a particle. This went back and forth for a while until physicists finally "solved" the problem by declaring that the light is both a wave and a particle. Soon, it was discovered that the electron, an obvious particle, could exhibit wave-like diffraction behavior. After a while, a satisfactory theory was developed, and we now know that any particle is a wave and any wave is a particle. And since aggregated waves represent a wave, a collection of particles is a wave too. Consequently, everything is made out of waves. You, me, this book, and all the objects around us have their wave representation.

Incredibly, mathematicians had a very similar revelation, but we did it a century earlier - of course. It all started with Joseph Fourier who studied the mathematical theory of heat conduction. He established and solved the partial differential equation governing heat diffusion. This accomplishment was worthy of immortality on its own, but the way he did it is even more important. While solving these equations, Fourier introduced a concept of trigonometric polynomials (series), and actually demonstrated that trigonometric functions constitute a natural way to characterize (essentially) any function. This was a strange development since by that time scientists as well as mathematicians associated trigonometric functions with waves.

But no. Fourier showed that trig functions are not only related to waves, but that essentially any function, whether it is "wave-like" or not, can be represented as a combination of Sines and Cosines. This powerful idea gave rise to a whole new branch of mathematics (Fourier analysis), and one cannot help but notice the parallel between the physics and the mathematics: *Wave-particle duality claims that any particle is a wave and Fourier analysis claims that any function is a trigonometric series (that is, a wave).*

2.3 The Logic and the Logic Circuits

> *A joke*: A mathematician, a physicist, and a mathematical logician fall asleep while traveling through the European countryside. When they wake up, they see a black sheep on a hill. "Look," exclaims the physicist. "In this country, all sheep are black." "But no, you cannot claim that," says the mathematician. "All you can say is that in this country there is one black sheep." "No, no, no," the logician complains. "That is a preposterous claim. All you can say is that in this country there is one sheep that is black on one of its sides."

2.3.1 *The Logic*

By now the reader is probably getting the gist of the game we mathematicians play. Unlike scientists, who practice reverse engineering - that is, they observe nature and go backwards trying to reconstruct its laws - we start with the laws and then develop a theory from them. The rules of our game are simple: Start with any set of laws (axioms) and then using only logic and pure reason deduce as many properties as you can. A theory is considered better if one can produce a richer, more elaborate web of interesting, unexpected properties. This is a very important point. These newly developed properties (theorems) need not make any practical sense, but they must be correct and, if possible, interesting and unexpected. Nobody likes to see a theorem that is a trivial consequence of an axiom.

Officially, we are free to choose any set of axioms and build a branch of mathematics out of it, but the reality is a bit more complicated and things not as simple. It turns out that it is very hard to pick a good set of axioms. Think of board games. There are a myriad of them, but only a few timeless classics like chess and Go. These two stand out since they have endured, and they have lasted because they have simple rules and yet they are able to produce extremely exciting games with countless interesting positions. In mathematical terms, their axioms are good because they are simple and still able to produce numerous interesting theorems.

On the other hand, the game of tic-tac-toe is considered trivial. The axioms are simple but so is the game. We do not see many interesting and unexpected positions with tic-tac-toe. Then there are games that are too complicated and inconsistent. Probably the best example would be the U.S. Constitution and the set of laws derived from it. As much as the founding fathers tried to make things self-evident and clear, and as much as the whole situation resembles mathematics (axioms–articles, theorems–laws), the Constitution has its loopholes.[2] The whole set of laws that are derived from these axioms is, from a mathematical point of view, an utter mess, packed with contradictions.

Mathematicians face the same problem: If one starts with some, any, set of axioms, one will most likely end up with a theory that is either too simple or too complex. And the too complex theories will most likely be contradictory. One needs to be a genius, a true virtuoso, in order to perform this very delicate balancing act of choosing axioms that are not too simple and not too complex, but just right. Throughout history we have very few instances of successful axiomatizations. It is interesting to observe that many of these successful axiomatizations were influenced by the natural world and our experiences within it. Thus, the very foundations of mathematics are tainted by our feeble senses.

Allow me to elaborate.

2.3.2 *Euclid's Axioms of Geometry*[3]

Axiom 1. It is possible to draw a straight line from any point to any point.
Axiom 2. It is possible to produce a finite straight line continuously in a straight line.

[2] Kurt Gödel himself advocated one on his citizenship exam – apparently the Constitution does not forbid a dictatorship.

[3] In literature one often finds references to Euclid's postulates and not axioms. In this book I will use *axioms* when dealing with mathematics and *postulates* when dealing with physics.

Axiom 3. It is possible to describe a circle with any center and distance.
Axiom 4. All right angles are equal to one another.
Axiom 5. If a straight line falling on two straight lines makes the interior angle on the same side less than two right angles, the two straight lines, if produced indefinitely, meet on that side on which the angles are less than two right angles (that is, two nonparallel lines must eventually intercept).

Clearly these axioms are inspired by observations. One really needs to draw or see a circle in order to realize that it is determined by its center and its radius. Euclid's Fifth Axiom is an even better example, for one really needs to experiment a bit before realizing that indeed two parallel lines will never meet.

However, mathematicians are looking for any collection of "good" axioms, and in this quest, they are not limited to sensible ones. The story behind the fifth axiom fits nicely in our context. Namely, Russian mathematician Nikolay Lobachevski showed that one can replace Euclid's Fifth Axiom and assume that two "parallel" lines could asymptotically meet.[4] He showed that this strange assumption does not create any contradiction. All we get is another game (theory) that produces a wealth of nice theorems. The fact that these results do not fit reality was not an issue.

It is important to pause here and observe that mathematicians had produced strange results before, but this particular case is special. Why? Because this was the first time that mathematicians deliberately introduced bizarre axioms, ones that clearly go against common sense. Nevertheless, the theory seemed correct, for it did not create contradictions, and research continued. The results did not make any practical sense but this did not stop Lobachevski, who claimed: "There is no branch of mathematics, however abstract, which may not some day be applied to phenomena of the real world."

And he was right. We have already mentioned that his "hyperbolic geometry" is intimately connected to the theory of relativity. But how did he know? Einstein developed relativity almost a hundred years later and, unlike many other theories, the theory of relativity did not have many predecessors so there were absolutely no hints that this bizarre geometry could be of any use. But it was!

Now, back to the point I am trying to make. It is true that choosing the "right" axioms is a very difficult job and it does not happen very often. But once it does happen, once we do get "good" axioms, mathematicians rejoice. We get a new game to play and all we need to do is to apply the rules of logic and pure reason and derive as many theorems as possible.

Enter George Boole, who dared to ask the obvious question: What *are* the rules of logic and pure reason? Not surprisingly, mathematicians assumed that the rules of pure reason are so self-evident and fundamental that it was almost blasphemy to even ask such a question. Trying to mathematicize logic seemed paradoxical. For you see, Boole was trying to mathematicize mathematics itself, and he was walking

[4] Much is written about the Lobachevski–Bolyai–Gauss saga, and who did what first. Nevertheless, all the evidence indicates that it was Lobachevski who published his work first.

on thin ice here. He proposed to analyze and derive the rules of logic and pure reason by using the very same logic and pure reason.

Mainstream mathematics did not look favorably on this idea (and we can only imagine what the scientists thought about it). This was around the mid-nineteenth century, and many math problems were still very much inspired by physics, geometry, calculus, analysis, and the like. These were the "important" open areas of research, and conventional mathematicians found Boole's ideas marginal, and his work an unnecessary digression. Nevertheless, George worked hard and managed to axiomatize logic and put the proper foundations in the very essence of mathematics.

From his axioms, he derived the whole set of rules and was able to show that essentially any statement, no matter how complicated or convoluted, can be reduced to the initial building blocks – the axioms. We will not go into detail here, but we present the following few brainteasers to illustrate the usefulness of formal logic.

Take the statement: *John loves cats and John loves dogs*. What is the negation of this statement?

What is the negation of the statement: *Everybody loves Raymond*?

Consider the statement: *If the earth is flat, then the Pope is not Catholic.* Is this a true statement or not?

Suppose you want to disprove the following statement: *If function f(x) is differentiable, then it is both continuous and integrable.* What should you prove?

Answers: *John does not like cats or John does not like dogs. Somebody does not like Raymond. True. You need to find a differentiable function which is either discontinuous or nonintegrable.*

Each of these brainteasers is easily answered once you break them down using Boolean algebra and the tools of formal logic. Boole's work proved essential to mathematicians. The field he established is called mathematical logic and it is considered one of the purest branches of mathematics. Logic is our backbone since it ensures rigor and provides the fundamental tools for the rest of us, in a way similar to how "regular" mathematics ensures rigor and provides the fundamental tools for physicists.

2.3.3 Logic Circuits

Boolean algebra was not an instant success. It was pretty much ignored by the mainstream mathematics and, aside from a few philosophers and logicians, very few people even knew it existed. It was Claude Shannon who observed that Boolean algebra supplies tools that are tailor-made for electrical engineers and computer scientists. The Boolean concepts *true* and *false*, which were represented by 1 and 0, are easily mimicked by electronic on–off switches. Mathematics derived a century earlier provided invaluable help in computer design.

Nowadays we often associate logical machines with computers and electronic gadgets, so much so that we do not even recognize the scale of the gap that was bridged here. These "logical" machines – computers, calculators, remote controls, and such – are not logical at all. They are just machines. In their essence, they are no different from sewing machines or vacuum cleaners. They are physical devices, designed to trick and manipulate the laws of physics in a way that is useful to humans; nothing more and nothing less. A cellular phone is not logical, nor is your digital watch. The only "logic" in there comes from the algebra derived from George Boole.

But this should come as a great surprise, a shock actually. Who would imagine that logic, this purest of the pure theories, would become the very backbone of modern industry? We can only imagine what the great Greek philosophers would say after learning that "logic circuit" actually refers to an engineering gadget. Socrates must be turning in his grave. Thus, the tale of Boolean logic has brought to the surface yet another unforeseen incursion of mathematics into the word of nature, but this time we have an additional twist. Unlike trigonometry or conic curves, which were well-established and useful mathematical theories on their own, without any help from scientists or engineers, Boolean algebra was not. There were *no* indications that this theory would be of any use to anybody, except to a very small circle of pure mathematicians. In many ways, Boolean algebra owes its reputation to this profound "coincidence," this strange twist, and its applicability to the world of technology.

2.4 Flowers and Very Spiky Curves

2.4.1 Spiky Curves

One simple way to visualize a curve is to think of a wiggly line that a pencil leaves on a piece of paper. Mathematicians have been studying specific curves for millennia, but it was not until Descartes's invention of the coordinate system that we were able to connect the visual, geometric aspect of curves with analytic expressions, that is, their formulas. For example, our good old friend the ellipse is nothing more than the collection of points $P = (x, y)$, satisfying the equation $\frac{x^2}{a^2}+\frac{y^2}{b^2}=1$. This fundamental tool, the Cartesian coordinate system, coupled with the invention of calculus, now allows us to address scores of intuitive geometric ideas like continuity, tangents, and curvature in a more precise, mathematical way.

But getting here was not an easy ride. Take, for example, a simple, almost trivial property like continuity. This should not be hard to address, since it is very simple to spot a discontinuity on a curve. All you need is to look for is a gap, a place where the pencil was lifted. However, the proper, mathematical way to define continuity is not so simple. Judge for yourself:

> We say that a function $f(x)$ is continuous at the point x_0 if for every $\varepsilon > 0$ there exists $\delta > 0$ such that $|x - x_0| < \delta \Rightarrow |f(x) - f(x_0)| < \varepsilon$. Otherwise, f(x) is not continuous at the point x_0.

Similar and rather complicated definitions were introduced for the slope of a tangent line (derivative), curvature (second derivative), length of the curve, and many other characteristics. With proper definitions, mathematicians went on proving theorems, most of which went along the Greek (Euclid) line. Slowly but steadily, they proved and generalized some important but intuitively expected facts (*the obvious math*). They

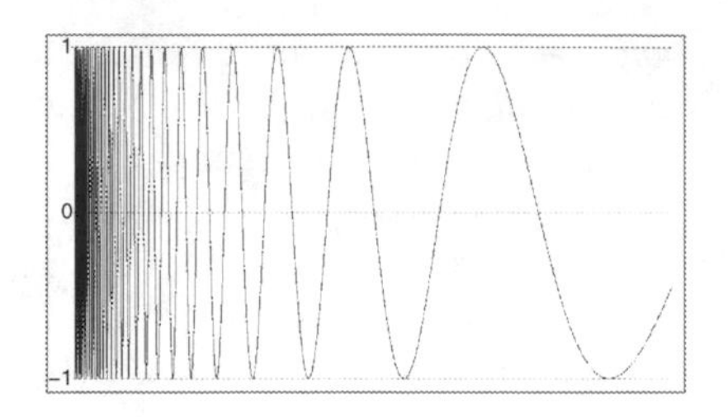

Figure 2.3 Plot for Sin($1/t$).

derived many useful formulas and calculated areas, curvatures, and maximums for numerous interesting and important functions (*the just-right math*).

But, as I hope the reader has come to expect, there were a few daredevils who opted for the unusual, bizarre mathematics - pushing the formulas and leaving paper and pencil behind. For example, one can easily construct a nice curve that is continuous, lives within the unit square, and is, however, *infinitely long* (just look at the left-hand end of $f(t)=Sin(1/t)$ in Figure 2.3).

Also fairly easily, one can construct a function that has a gap (discontinuity) at every rational point (every fraction) but that is continuous in all other points. Now, this is a weird creature. The fractions are *dense,* and this means that in any interval, no matter how small, the curve has infinitely many gaps; yet for the most of the points the function is continuous (the function is defined as $f(t) = 1/m$ if $t = n/m$ and $f(t) = 0$ otherwise).

Giuseppe Peano's[5] space-filling curve was another shocker. Peano constructed a continuous, one-to-one curve whose graph is not the usual "wiggly line" but a whole unit square. Clearly, neither the curve with infinitely many gaps nor the space-filling curve could be plotted. The only way to study them is to rely on mathematics.

Probably the most disturbing curve is one constructed by Karl Weierstrass. His curve is continuous but does not have a derivative at any point. Mathematicians did not welcome this development. In fact, Charles Hermite decried "this dreadful plague of continuous nowhere differentiable curves," while Henri Poincaré objected: "Yesterday, if a new function was invented it was to serve some practical end; today they are specially invented only to show up the arguments of our fathers, and they will never have any other use."

Why was Weierstrass's curve so disturbing? To begin with, we have to understand that the nineteenth century was a great conciliatory period between physics and mathematics. Many big, open mathematical questions were inspired by thermodynamics and mechanics. In this milieu, a continuous function was associated with a moving object. And with good reason. If we think of the variable t as time, then $f(t)$ gives the position of an object at time t. Clearly, $f(t)$ must be continuous for otherwise one would have an object that, at some point(s), traveled a fixed distance

[5] Please do not pronounce his name as Piano.

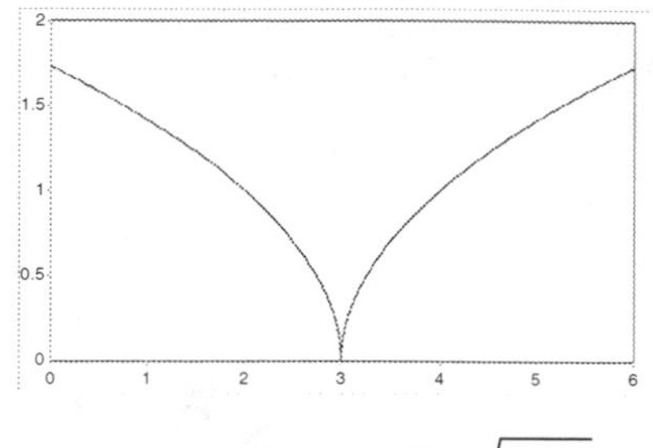

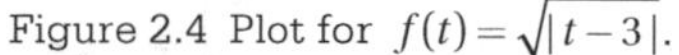
Figure 2.4 Plot for $f(t) = \sqrt{|t-3|}$.

in 0 seconds. Also, ever since Newton and Leibniz invented calculus, the derivative of the function $f(t)$ at time t naturally corresponds to the velocity of the object.

It is very easy to construct a continuous function that does not have a derivative at *some* point (one example is $f(t) = \sqrt{|t-3|}$), but these are easily understood instances for which we have a physical interpretation. We can graph this curve, as in Figure 2.4, and the graph reveals that at timepoint 3 the "object" bounced off the floor and instantaneously changed direction. Clearly, our intuition of velocity and the tangent line at that spiky point breaks down and we just declare that at timepoint 3 the object does not have velocity (that is, the function $f(t)$ does not have a derivative).

Now you see the problem that Weierstrass's function $f(t)$ poses. Since it is a continuous function, it represents the path of a moving object, but since $f(t)$ does not have a derivative at any point this moving object does not have velocity at any point in time. But how does it move then?

2.4.2 *The Flowers*

Oblivious to all this, and around the same time, an unlikely figure entered the picture. The person in question is Robert Brown and his tale deserves a few extra lines.

The time was the early nineteenth century. The microscope had become relatively accessible and many new discoveries of organisms, not readily seen by the naked eye, were being made. Bacteria, amoebas, spores.... It seemed that something new was discovered in almost any body of liquid one looked at. Brown decided to take a closer look at the behavior of flowers' pollen particles. The consensus of the day was that pollen, unlike sperm, is not alive and consequently should not move. But move it did.

Brown observed that pollen from the ragged robin flower, placed on a water surface, would exhibit the unpredictable zig-zag, stop-and-go motion commonly associated with living things, such as bacteria and sperm. And he was very methodical. He worked with old pollen and with fresh pollen, he baked the pollen at different temperatures in order to "kill it," but the motion persisted. He even experimented with volcanic ash (that could not possibly be alive) and he still observed this strange *Brownian* motion, shown in Figure 2.5.

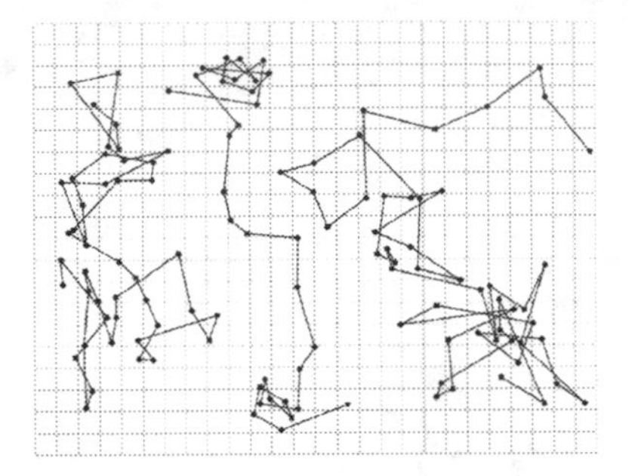

Figure 2.5 Brownian motion.
Note: Reproduced from the book of Perrin, *Les Atomes*: Three tracings of the motion of colloidal particles of radius 0.53 μm, as seen under the microscope, are displayed. Successive positions every 30 seconds are joined by straight line segments (the mesh size is 3.2 μm).

A good 70 years later a Frenchman, Louis Bachelier, studied a completely unrelated problem: the fluctuations of a stock market. He postulated a few axioms, very commonsense assumptions that ought to be satisfied for this particular fluctuation. For example, the past cannot influence the future, or in concrete terms, yesterday's price change cannot predict today's price change, for if it could, one could get extremely rich very easily. He derived a mathematical model for this motion but to his misfortune he was a mathematician, and as such he shouldn't have "dirtied" his hands with applied problems like the stock market.

His doctoral advisor, the famous Poincaré (yes, the same Poincaré who abhorred Weierstrass's function), was not overly happy with the thesis in which this model was presented. Bachelier received a grade of "honorable" and in order to be considered for an academic post one needed "very honorable" mention. A few years later, another familiar name entered the picture. In a paper appearing at the same time as his papers on the photoelectric effect and the special theory of relativity (but much less publicized) Einstein derived the mathematical description of Brownian motion and used it to finally prove the atomic model.[6]

The culminating figure in this saga is Norbert Wiener, who developed a powerful mathematical apparatus and once and for all established the mathematical theory behind this motion. He showed that starting with axioms very similar to Bachelier's, one could show that there exists a way to mathematically model Brownian motion. This model is called the Wiener process. In this model, mathematicians received their new toy to play with, and a situation similar to one described at the beginning of this chapter arose again. Namely, mathematicians had explicit formulas that describe this geometric object, and consequently they could ask all kinds of interesting questions and hope that the equations would deliver the answers. And they did.

[6] Much as with Euclid's Fifth Axiom, here too we have some controversy. Who did what and when is not exactly clear. The reported extent of Bachelier's contribution varies dramatically, depending on the sources one reads.

2.4.3 The Finale

So far, these two tales of flowers and of spiky curves do not have much in common. And here it comes. The pollen particle moves in time. So, if we think of $f(t)$ as the position of a particle at the time t, we have a curve. Next, since we have the proper mathematical model for this curve, we can derive its mathematical properties. For example, one of the axioms of Brownian motion postulates that this motion, if plotted against time, must be a continuous function. This is a fundamental assumption, for one cannot accept a discontinuity here. It would mean that a particle jumped X inches in *zero* seconds, resulting in infinite velocity. But, alas, it can be proven that if one "observes" the Wiener process (aka Brownian motion) for any period of time (say a second), one would observe a curve that is infinitely long. Thus, although we forbade infinite velocity (by assuming the Continuity Axiom), mathematics played a trick on us, for it allowed an infinitely long path in finite time – thus infinite velocity.

But, of course, the grand finale is related to the following result: It can be proven that the curve traced by the Wiener process (just like Weierstrass's function) is *not differentiable at any point.* That is, we cannot properly define the velocity for any single time instance t, and yet, the particle obviously moves. Thus, one of the most bizarre mathematical curves, one believed to have no motivation except to test the boundaries of mathematics turns out to be intimately related to the very motion of molecules and the stock market.

We should observe here that Weierstrass's function was not a well-established or useful mathematical concept, so in some sense the story of its recognition in physical reality is similar to that of Boolean algebra. In both cases we see an obscure mathematical result playing an important role in some unrelated natural phenomenon. But, the tale of the "spiky curve" offers yet another twist that intensifies this notion of mysterious mathematical prophesies: Boolean algebra might have been obscure and recognized only by a handful of purists, but it was *designed* to be helpful to a handful of purists. Weierstrass's function was deliberately created *not to be useful.* It was designed *only to show up the arguments of our fathers.* Yet, somehow, the mathematical oracle entangled this twisted creation into the very fabric of the nature: the movement of the atoms.

2.5 Conclusion

I hope the reader will agree that these stories are compelling. Example after example points toward this strange mathematical prophecy. Somehow, mathematics seems to have a mind of its own – and a rather independent mind – which sometimes behaves contrary to expectations. Mathematics predicts the laws of nature even though its creators, mathematicians, did not wish to do so. Quite the contrary, mathematicians deliberately avoided experiments and concentrated only on logic and pure reason. Yet, somehow, inadvertently, century after century they kept delivering mathematics that is in perfect harmony with the laws of physics.

The Greeks studied curves in connection to a dead-end problem (doubling the cube), and a few thousand years later these curves played an essential role in describing the earth's orbit around the sun. Trigonometric functions, introduced to help us solve the geometrical problems of triangles, somehow proved to be invaluable tools for describing a completely unrelated physical phenomenon: waves. And the list goes on. We have examined here only four instances (actually quite a few more if one reads between the lines), but there are literally hundreds, if not thousands more. Nevertheless, we did not solve the mystery. We did not pinpoint the answer to: *Why? How?*

Maybe if we dig deeper, if we examine some of these mathematical intrusions into the world of science and technology more carefully, maybe we will spot the *eureka* moment, the very mark where mathematics imposed itself. Let us try.

2.5.1 Checking the Math

Let us revisit the discussion of music and angles and the unexpected connection between *trigonometric functions* and *oscillations (or waves).*

> I urge the reader to be patient. The following equations need not be understood in detail. It is enough to skim through them in order to get the "feel."

The physics is straightforward. The simple harmonic oscillator that has no driving force and no friction has the net force described by,

$$F(t) = -ks(t)$$

where $s(t)$ is the position at time t and k is a positive constant. Newton's second law says that $F(t) = ma(t)$. Putting the two together, we get the following:

$$F(t) = ma(t) = -ks(t) = F(t).$$

By definition, the acceleration, $a(t)$, is equal to the second derivative of $s(t)$, and by making the substitution we get the following differential equation:

$$s''(t) = -s(t)k/m.$$

The general solution for this differential equation is $s(t) = e^{\lambda t}$, which yields

$$\lambda^2 e^{\lambda t} = -\frac{k}{m} e^{\lambda t}.$$

So far so good. The above equation demands that $\lambda^2 = -k/m$, and here is where the first problem arises. In order to solve for λ, one must take the *square root of a negative number*, which yields $\lambda = i\sqrt{k/m}$, where *"i"* is the good old imaginary unit. Next, we use the notation $\omega^2 = k/m$ and Euler's formula $e^{ib} = \mathrm{Cos}(b) + i\,\mathrm{Sin}(b)$ to get the solution

$$s(t) = \mathrm{Cos}(t\omega) + i\,\mathrm{Sin}(t\omega).$$

Confused? Of course you are. Everybody should be. We got our mathematically derived formula and the *oscillations* to indeed follow the trigonometric functions, but the intuition is *completely lost.* We used the imaginary number i, which is fictional and does not refer to any geometric (physical) reality. The solution derived above is correct. One could easily check by plugging it back into the original equation. But what is the meaning of all this? What happened with the triangle and the original definition of $\mathrm{Sin}(x)$?

Trigonometric functions appeared in the above computation only because the second derivative of Sin(x) is again Sin(x) (with a minus attached). This fact is easy to check, but it is also a fact that does not reveal any inside information, nor does it explain the connection between waves and triangles. And let us not forget that the Greeks and the Arabs, the people behind Sin(x), had absolutely no way of knowing this strange property of Sin(x) related to derivatives. Infinitesimal calculus was introduced at least a thousand years later.

And this is very typical. For centuries now, it has been clear that while mathematics helps scientists, most of the time this help comes in mysterious, completely nonintuitive, matter-of-fact ways. It all echoes the arguments we used to have with our parents:

Can I have a cookie?
No, we're eating dinner soon.
What does dinner have to do with me having a cookie?
It will spoil your appetite.
So what if I spoil my appetite?
It's not good for you.
Why isn't it good for me?
Because I said so!

And this is exactly what happens if one asks too many questions and tries to understand why math helps scientists. It helps because it does. Nobody knows "why" the second derivative of Sin(x) is again Sin(x), but this fact is the reason we have trigonometric functions in the solution of the harmonic oscillator equation. There is nothing more to it. Or is there?

2.5.2 *Experiment vs Philosophy*

Throughout history, there have been many who tried to use logic and pure reason to explain our world and to derive its laws. But they have all failed miserably. All of them, except mathematicians. Indeed, for millennia, slowly but steadily, we have seen new scientific discoveries appearing in a systematic and orderly succession. New materials and construction techniques have been introduced, new crops and farming methods, new weapons, new ships. Although many of these developments have been closely related to science, scientists did not produce them. Farmers, engineers, generals, and sailors carried the torch. In fact, one of the most embarrassing chapters in human history is the history of science.

Scientists, bless them, have tried very hard. Using logic and reason, they have asked all the right questions: What is life? How does the human body work? What is the shape of the earth and what is its relation to the sun? But the answers they offered, well, they have been absurd. *Earth as the center of the universe*, a *flat Earth*, the *Sorcerer's Stone*, and *an Elixir of Life* are just some of a myriad of ridiculous notions the ancient scientists cooked up. They were not just "a bit off" or "somewhat right." No, they were completely wrong. In fact, fourteenth-century "scientists" had no better perception of the world around them than the ancient Mesopotamians five *thousand* years earlier. In many ways ancient science did not advance at all. It receded.

One could try to rationalize and downplay these failures, but one would have very hard time explaining the 5,000-year waiting period for something as simple as the need to boil drinking water or to wash our hands. I do not see how generations of "doctors" could miss the obvious connection between hygiene and health, or the evident disinfecting properties of alcohol. How, after several thousand years of engineering, did "scientists" not figure out that that potential energy is related to kinetic energy and that the latter depends on the square of velocity? Engineers knew all of this very well. They had developed comprehensive collections of "rules of thumb," numerous techniques, catalogs, and tables. What they had not done was discover the laws of nature. That was not their job. Engineers are here to build things and not to philosophize. But how could scientists have missed all this?

In order to solve this puzzle, we need to examine the scientists who were successful. What was their secret? The answer turns out to be quite simple: *They experimented.* Ancient scientists were philosophers, and consequently, they preferred reasoning and shunned experimentation. And sure, if one looks around one can "reasonably" conclude that the earth is flat and it is "logical" to conclude that the sun goes around the earth. Without experiments to confirm or refute these theories, it is not surprising that the whole process went astray. Had doctors actually gone among their patients and observed their conditions in a systematic way, I am confident that within a generation or two we would have known that hygiene is key, and everybody would have been boiling water. Had they done this a few thousand years ago (and nothing had prevented them from doing so), today's world would be a profoundly different one.

But it was not until the seventeenth century that Galileo (physics) and later Lavoisier (chemistry) finally put matters into the right perspective. The experiment and the validation of a theory become the centerpiece of a new dogma. This new breed of scientists insisted on rolling up their sleeves, putting on their boots, and collecting specimens and conducting experiments. The "philosopher-scientists" were abolished and enlightenment had begun. In just 200 years, this approach produced more relevant science than the previous 5,000 years combined. In fact, if we rule out Archimedes, it is next to impossible to find any law of physics discovered in the 5,000-year period between the Pyramids and Galileo.

And how do our mathematicians fit into this saga? They were the very champions of *reasoning without experimenting.* In fact, they went even further than the philosopher-scientists, for they did not even bother to observe nature, nor did they try to answer any fundamental questions. They just played their game, focusing on axioms; they used logic and pure reason and derived as many theorems as possible. Thus, one would expect that their fate would be even worse than that of the philosopher-scientists. But we know it wasn't.

It turns out that these two extremes, one of modern science and the other of ancient mathematics, complemented each other perfectly. Modern scientists disregarded philosophy and focused on observations and experiments, and, in the process, they discovered the laws of physics which are in ideal agreement with mathematics. Mathematicians used the completely opposite approach. They ignored experiments and focused solely on logic and reason and, in the process, they derived mathematics that is in perfect harmony with modern science and laws of nature. Spooky, isn't it?

3 On Modern Physics

By now, there should not be any doubt left about this spooky action of mathematics. This cannot be just a coincidence. There must be some kind of a process, or a design even, that governs these bizarre and frequent occurrences, for how else to explain all this? I hope the reader agrees: We cannot just brush this off. There must be a logical explanation. Our attempts, so far, to carefully examine the mathematics behind some of these instances have shed very little light. If anything, they have deepened the mystery. But what about physics? Maybe if we look at the problem from the scientists' (i.e., physicists') angle, we can gain some clarity, some glimpse of this process – this mechanism that forces mathematics onto science.

> *But first a joke:* How does a physicist demonstrate that all odd numbers are primes? By an experiment (of course): 1? odd and prime, check; 3? odd and prime check; 5? check; 7? check; 9? ... well, not a prime, but let us continue; 11? check; 13? check. Conclusion: All odd numbers are primes and 9 is an experimental error.

3.1 Planck and Quanta

Max Planck is considered the father of quantum mechanics, for it is he who first postulated that energy cannot be infinitely small, that there must exist a minimum amount of energy, called quantum. Planck discovered this phenomenon in a rather unusual way. It was a combination of experimental results (performed by his predecessors), some grueling theoretical, mathematical work, and a little luck. Lots of trial and error and very little intuition were involved. This story is extremely important and I would strongly encourage the reader to be brave and face the equations below; even if much detail is lost, it is worth investing a few extra minutes on these formulas. They tell an amazing tale.

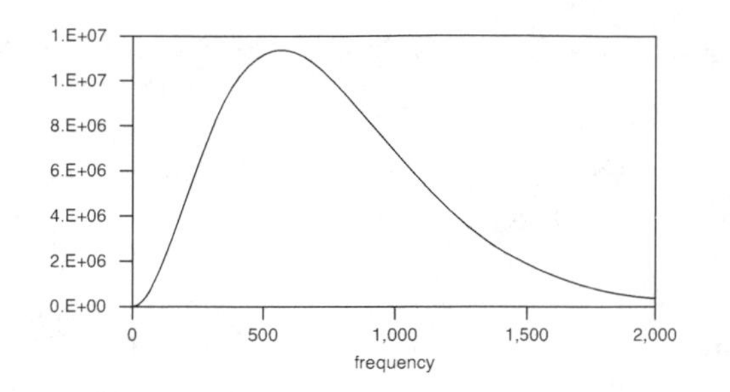

Figure 3.1 Black-body radiation. The *X*-axis is the frequency, the *Y*-axis is the amount of energy, and the temperature is fixed. One can see that the highest intensity of light is one with a frequency of 600, and light with a frequency of 2,000 has almost no intensity.

It all started with one of the very last, great unsolved problems of the nineteenth century: the black-body radiation problem. The problem, in a nutshell, is related to the well-known phenomenon in which a heated object gets brighter as it gets hotter. We have all experienced this process. An electric stove turns from red to orange as it gets hotter, and in reverse, the campfire logs glow deeper red as they cool down.

As the experiment goes on, it becomes clear that there exists a relationship between an object's temperature and the color of the light it emits. For each temperature, depending on the material of course, there exists a dominant color that is accompanied by a whole spectra of less intense colors. The intensity of different colors is easy to measure and consequently a very precise curve has been derived (Figure 3.1).

The problem was that physicists could not derive the formula for the above (experimental) curve. They could not produce an equation that would be in agreement with both the laws of physics and the experimental results. Classical physics, coupled with modern (at the time) statistical mechanics derived by Boltzmann, was very successful in many other situations but fell short in this last example. The idea behind the molecular model of heat is to relate molecules' kinetic energy with temperature. In a nutshell, the faster the molecules move, the greater their kinetic energy. This implies more molecular collisions, which in turn yields higher temperature. The heat is proportional to the energy one gets from these tiny (but numerous) molecular collisions. Boltzmann's formula "counts" the number of instances (molecules) at energy level E and it is given by $P(E) = Ae^{-E/kT}$.

Physicists successfully applied this formula in numerous cases, all of which had experimental confirmation as well as theoretical backing. And it was a great success. For example, using this approach one can easily compute the average energy E at temperature T. By definition the average energy is given by the following formula:

$$\bar{E} = \int EP(E)dE \,/ \int P(E)dE.$$

By substituting $P(E) = Ae^{-E/kT}$ and applying simple calculus formulas, one can derive

$$\overline{E} = kT.$$

Thus, Boltzmann's formula, with some help from mathematics, produced this classical result, one that had been experimentally confirmed in numerous instances. It simply states that the average energy emitted from a heated object is proportional to temperature T. The constant k relates to different materials.

This was the obvious starting point for the black-body problem. Since a heated body radiates light that has different frequencies at different temperatures, one needs to connect the thermal energy formula with the light and wave energy. Classical physicists understood waves very well and could connect the frequency, the amplitude, and the energy. Thus, after some work, the Rayleigh–Jeans formula was delivered:

$$E(\nu) = \frac{8\pi\nu^2}{c^3} kT,$$

where ν stands for frequency and T for temperature. The remaining symbols are constants.

The formula failed terribly. It predicts that intensity of heat increases quadratically with the frequency (depending on ν^2), which would imply that energy tends to infinity for higher frequencies. This is in obvious disagreement with the experimental curve (Figure 3.1) where the intensity tends to zero for higher frequencies. It also contradicts common sense.[1] A piece of metal heated to 800° Celsius will glow red. Thus the majority of energy emitted will be in red, low-frequency spectra, with a minimal amount being observed at very high frequency (bright white).

The next attempt to salvage the situation came from Wilhelm Wien. He produced the formula that replaced the kT part with more suitable $k\beta\nu e^{-\beta\nu/T}$. His formula,

$$E(\nu) = \frac{8\pi\nu^2}{c^3} \frac{k\beta\nu}{e^{\beta\nu/T}},$$

matched the experiments much better, in particular for large frequencies ν. However, it failed to do the same for the smaller frequencies. To make the story more bizarre, the Rayleigh–Jeans formula, which failed terribly for large frequencies, worked very well for small frequencies.

Enter Max Planck. With a simple modification, just adding "–1" in the denominator, he produced the "correct" formula:

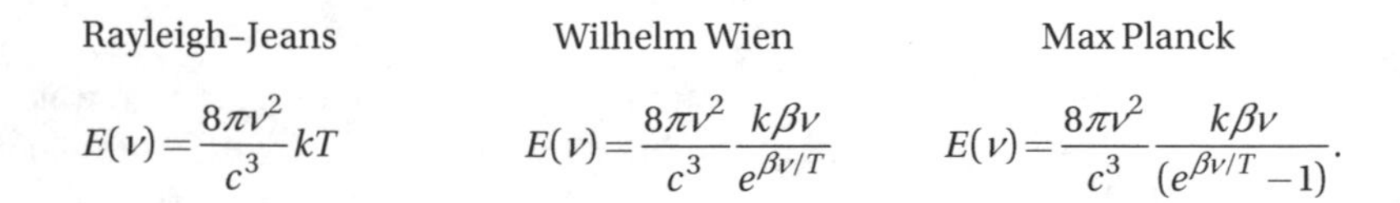

Rayleigh–Jeans	Wilhelm Wien	Max Planck
$E(\nu) = \frac{8\pi\nu^2}{c^3} kT$	$E(\nu) = \frac{8\pi\nu^2}{c^3} \frac{k\beta\nu}{e^{\beta\nu/T}}$	$E(\nu) = \frac{8\pi\nu^2}{c^3} \frac{k\beta\nu}{(e^{\beta\nu/T} - 1)}.$

[1] This is the source of the well-known phrase "ultraviolet catastrophe."

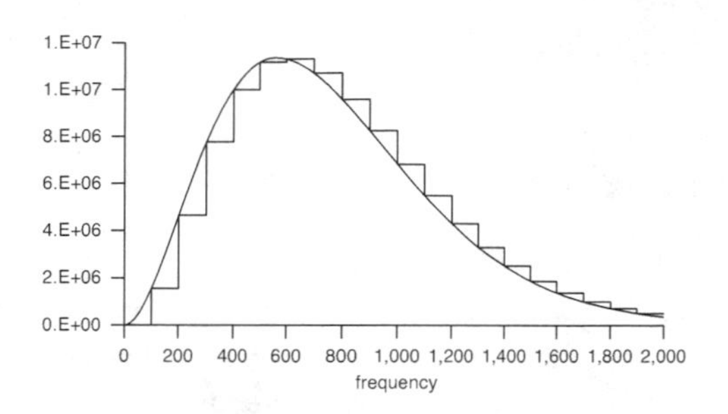

Figure 3.2 Riemann integral. A curve is approximated by a step function whose area is easily computed as the sum of rectangle areas with base δ and height $f(i\delta)$.

This formula worked like a charm. First, it matched the experiments. This goes for large ν's as well as for small ν's. Moreover, it provided the mathematical connection between the two previous formulas. Namely, by taking the limits (see the Appendix) one could easily show that for small ν's Planck's formula behaves like Rayleigh–Jeans's formula (i.e., it behaves like ν^2), while for large ν's it reduces to Wien's formula.

The problem, of course, was that one cannot just "*stick* (-1) *into formula*". One must have some justification. And Max had none. He worked very hard and still could not find the "reasons" for this "obviously" correct equation. "Massaging" the mathematics for several weeks ("the worst weeks of my life," he allegedly said) he found the solution, which laid the foundation for quantum mechanics. A theory that was in contradiction with classical mechanics, as well as calculus. See for yourself.

> *A plea from the author*: Please do not get discouraged with the formulas. Even if the details are "murky," the main point will be clear, and your work will be rewarded. Trust me on this.

Remember the definition for average energy $\bar{E} = \int EP(E)dE \,/ \int P(E)dE$ and Boltzmann's formula $P(E) = Ae^{-E/kT}$. Simple calculus yielded the formula $\bar{E} = kT$. Unfortunately, this "kT" is the same "kT" that appears in the Rayleigh–Jeans formula:

$$E(\nu) = \frac{8\pi\nu^2}{c^3} kT,$$

which was a dead end. So, Planck tried something else. He rolled up his sleeves and integrated the function, but he did not follow the usual way. He used the rectangle approximation of the area under the curve (Figure 3.2). Typically, there is nothing wrong with this approach, other than the waste of time.

Formally, $\int f(x)dx = \lim_{\delta \to 0} \sum_{i=0}^{\infty} \delta f(i\delta)$, where δ is the width of the rectangles and $f(i\delta)$ is the height. In layman's terms: If we take "millions" of these rectangles, each with an extremely small width (small δ), our approximation of the area becomes the actual area (thus the *lim* part). Of course, one does not need to compute these limits for every function of interest. That is why we have calculus. It provides the shortcuts and formulas designed to facilitate this process (table of integrals, substitutions, integration by parts, etc.).

This is a good spot to pause and reflect on the frustration Planck must have faced. Like a desperate engineer who could not pinpoint a machine's failure, Planck started dismantling his "engine," frantically hoping that somewhere, along the line, a revelation might present itself. He needed some "miracle" that would justify his "obviously" correct formula. This is the reason he decided to dismantle the standard integration formulas. Essentially, he pretended he did not know calculus and approached the problem from scratch. Thus, in order to compute $\overline{E}$, he did not integrate, but used the cumbersome summation formula instead:

$$\overline{E} = \frac{\int_0^\infty AEe^{\frac{-E}{kT}}\,dE}{\int_0^\infty Ae^{\frac{-E}{kT}}\,dE} = \lim_{\delta\to 0}\frac{\sum_{i=0}^{\infty} i\delta^2 e^{\frac{-i\delta}{kT}}}{\sum_{i=0}^{\infty}\delta e^{\frac{-i\delta}{kT}}}.$$

Next, with some simple algebra (curious readers can find the computation in the Appendix of this book), the above limits reduce to

$$\overline{E} = \frac{\int_0^\infty AEe^{\frac{-E}{kT}}\,dE}{\int_0^\infty Ae^{\frac{-E}{kT}}\,dE} = \lim_{\delta\to 0}\frac{\delta}{e^{\frac{\delta}{kT}}-1},$$

which, after applying the L'Hospital rule, yield

$$\overline{E} = \frac{\int_0^\infty AEe^{\frac{-E}{kT}}\,dE}{\int_0^\infty Ae^{\frac{-E}{kT}}\,dE} = \lim_{\delta\to 0}\frac{\delta}{e^{\frac{\delta}{kT}}-1} = kT.$$

And we are back where we started, stuck with $\overline{E} = kT$, which failed miserably. And we should be stuck, since solving the same problem, using a different approach, should yield the same result. Mathematics is very "fussy" about that. But frantic Plank noticed a curious little thing: "−1" appeared in the denominator. You see it? It looks very much like that "obviously" correct formula of his, if only ... If only, he could avoid taking that limit! But, of course, he could not do this. The very essence of infinitesimal calculus is buried in this idea of $\lim_{\delta\to 0}$. But what if...

What if one replaces the original limit $\delta \to 0$, and instead let $\delta \to h\nu$? In this case the mathematics yields

$$\overline{E} = \frac{\int_0^\infty AEe^{\frac{-E}{kT}}\,dE}{\int_0^\infty Ae^{\frac{-E}{kT}}\,dE} = \lim_{\delta\to h\nu}\frac{\delta}{e^{\frac{\delta}{kT}}-1} = \frac{k\beta\nu}{(e^{\frac{\beta\nu}{T}}-1)},$$

which is the desired formula (see the Appendix for details). Moreover, this cheating does not seem that bad since $h\nu$ is "almost" zero anyway (in fact h is a constant of the order 10^{-34}). But of course $\delta \to 0$ comes from the very definition of integral and one cannot arbitrarily replace it. Mathematics does not allow for that.

I hope the reader grasps the scale of this predicament. This mathematical trick opened the door to a very nice settlement. We got a formula that agrees with the experiments, it explains the two other theoretical approaches, and it can be derived from fundamental laws of physics. If only we could justify this strange, abrupt stopping of the limit. Well, one solution would be to *declare a new law of physics*, one that would forbid infinitely small energy and consequently would impose the rule that δ stays away from zero (remember that δ was the x-axis and we were integrating with respect to energy E).

This would solve the problem and it would be mathematically correct. But, as one could imagine, Planck himself was a bit skeptical. The "solution" came out of thin air, and although it proved correct, the whole idea of quanta continued to be uncomfortable for Planck. At the end of his paper, he wrote

> I hope, nevertheless, that we will find a solution using Newtonian dynamics.

And also:

> I tried immediately to weld the elementary quantum of action somehow in the framework of classical theory. But in the face of all such attempts this constant showed itself to be obdurate ... My futile attempts to put the elementary quantum of action into the classical theory continued for a number of years and they cost me a great deal of effort. M. Planck, *Scientific Autobiography, and Other Papers* (Philosophical Library New York, 1949, p. 44)

Disclaimer

The presentation above differs somewhat from the common accounts found in physics textbooks. I have adapted it so that it fits better with our narrative, and in the process I replaced some of the physicists' jargon with that of calculus. The tale I tell deviates slightly from Planck's actual account but the spirit (if not the details) has been preserved.

3.2 Einstein and Light

The story behind Einstein's special theory of relativity is well known and is part of intellectual folklore. I am confident that the reader has seen the theory's basic outline and is familiar with many aspects and anecdotes related to its development. This in itself confirms my belief that mathematicians have gotten the short end of the stick, so to speak. For millennia we have been developing our theory, most people are "strongly encouraged" to study it at school, scientists and engineers use it, and yet, somehow, nobody seems to know contemporary mathematicians.

Names like Einstein, Hawking, Curie, Watson, and Crick are all very familiar to most mathematicians, but ask the analogous question of a scientist, and most will have a hard time recognizing names like Banach, Hadamard, Gödel, or Littlewood. It does not seem fair, does it? But it fits very nicely with the thesis of this book. The theories that mathematicians develop typically become applicable decades, even centuries, after their authors have passed away, and by the time a general scientific audience becomes familiar with their names, they are not contemporary anymore.

But we digress. Let us go back to Einstein. His special theory of relativity is based on two postulates:

Postulate 1. The laws of physics are the same in any inertial (nonaccelerated) frame of reference.

Postulate 2. The speed of light is the same for all observers, no matter their relative velocities.

What mathematicians call axioms, physicists often call postulates. The big picture is similar, we both start with these "absolute truths" and derive the theory from them. The main difference is that physicists need to be right while mathematicians don't. Although this might sound like a joke, it is not. Our theories start with the following disclaimer: "Suppose that these axioms are true, then...." So in a sense we have an easier job because we "wash our hands" from the very beginning. We never claim that the axioms are correct, and consequently we do not worry if some of our results turn out to be in contradiction with physical reality. Physicists, on the other hand, have a much tougher job. First, they must derive their postulates, next they need to develop their theory, and then, as the final test, they need to experimentally confirm it.

As Einstein's postulates go, the first one seems straightforward. Anybody experiencing a long smooth ride in an elevator or on an airplane can testify that it is impossible to tell if you are moving or not. Einstein went one step further and claimed that it is impossible to construct any scientific experiment that would determine if your elevator is moving up or if the whole building is moving down. While you are in the elevator, you might believe that it is you that is moving and not the building, but at that very moment, while you are riding, you could never be sure. That is the essence of the first postulate.

The second postulate is a tricky one. It clearly goes against common sense and everyday experience. If you are catching a baseball thrown at you, and you do not have a glove, you might get bruised a bit but you would probably be OK. However, if someone is riding toward you in a speeding truck, and they throw a baseball at you, you are in serious trouble. Since the two velocities (the truck's and the thrower's) are combined, the ball is hurling toward you at a speed easily exceeding 100 mph. This and similar experiences clearly go against the second postulate, which states that in the case of light we do not add the velocities.

One does not need to be a genius to construct a paradoxical situation. Let us push things to the extreme and assume that a space ship is moving toward us with the velocity c (the speed of light). Suppose that the spaceship fires a "photon torpedo" at us. Something strange happens next. The ship's crew, due to the first postulate, must be able to see the torpedo leaving the ship. However, due to second postulate, for us on earth, the torpedo never left the ship, for if it did it would have traveled faster than light. So, which is true? Did the torpedo leave the ship or not?

Since the velocity is locked up by the second postulate ($c + v = c$), something else must give. And that something is "time" (and the distance s, since $v = s/t$). Using fairly simple, almost high school-level mathematics, one can derive most of the formulas presented in the special theory of relativity. An elegant geometric argument (we have all seen the triangles related to this story) yields a correction factor

$$\gamma = \frac{1}{\sqrt{1 - v^2 / c^2}}.$$

The genius of Einstein is not reflected in these formulas per se, but in the choice of the postulates. The situation is similar to the one described earlier with regard to Euclid's Fifth Axiom. There, Lobachevsky and Bolyai challenged commonly accepted Euclidian geometry and tweaked the axioms, only to derive an array of consistent but nonintuitive geometric theorems. Similarly, Einstein challenged Newton's mechanics, tweaked the "adding of the velocities" postulate, and, as a consequence we got arguably one of the most peculiar laws of physics (time slows down, mass increases, the space ships shrinks).

Strangely enough, it seems that Einstein was not fully aware of Michelson-Morley's experiment published a few years earlier,[2] in which they actually confirmed the first postulate. His genius is even more impressive if we factor this in, for it is almost unimaginable that someone could guess such a strange condition. How did he do it? What is the intuition behind the seemingly ridiculous assumption $c + v = c$?

There are many anecdotes and stories behind this intuition and I would like to offer a few of my own. To start with, light has a special meaning to us humans, for it is associated with our primary source of information, our eyesight. For all intents and purposes light is instantaneous. Lighting a candle in a cave illuminates the whole area "instantly" and so does the light bulb in a room. Thus, deep down we have this notion of light's infinite velocity. And infinity has these strange, but intuitive properties:

$$\text{Infinity} + v = \text{infinity}, \qquad \text{as well as,} \qquad \text{infinity} + \text{infinity} = \text{infinity}.$$

In other words, if we add something to infinity, the result is still infinity, which makes sense. If we replace *infinity* with the speed of light c, the above intuitive formulas translate to:

$$c + v = c \qquad \text{as well as} \qquad c + c = c,$$

which is essentially the second postulate. Einstein knew that the speed of light was not infinite of course, but, from a mathematical point of view, his postulate in many ways implies that it is. Namely, the proper mathematical definition of an infinite set

[2] Yet another controversy: Did Einstein know of Michelson-Morley's experiment or not? There are "indications" pointing to either side of the argument, but luckily for our purpose this issue is not important.

reduces to the requirement: "set is infinite if adding finitely many elements to it does not change its cardinality." In other words, if $X + v = X$ for any finite v, then X must be infinite. So, although we know that the speed of light is finite, the second postulate makes it de facto infinite.

The next obvious question would be: Why light? In theory, Einstein could have chosen some other velocity to be maximal. Why did he choose light? It turns out that if one allows velocities faster than light one could create many paradoxical situations. Consider the following argument: Imagine a world in which we could travel faster than light and imagine yourself and your prankster friend standing on the beach watching the sunset. Since he can move much faster than light, he decides to move very quickly and stand in front of you. What do you see? Amazingly, you see two of your friend simultaneously, since the light is still coming toward you from both sources, one in front of you and one next to you. But since you are not immune to a good joke, you decide to do the same to him, and now he sees two of you, but which one? There are four of you, all "watching" each other (in fact you could see yourself standing next to your friend).

Pushing this story to the extreme would create a nightmare for theoretical physicists. Later on, with the general theory of relativity, gravity was factored in, and here again, anything faster than speed of light would have created an impossible maze of paradoxes (if you travel faster than gravity, you do not feel it, etc.). Thus, it seems much safer and simpler to assume that nothing (not even a force field) could travel faster than light. And in physics, simpler is better.

3.3 Heisenberg and Uncertainty

Heisenberg's discovery is another jewel in the crown of modern physics. The essential statement of his (in)famous Uncertainty Principle is well known, but only a small group of specialists is familiar with the way the principle was discovered. The actual process behind the discovery fits very nicely with our quest and it would be a shame not to describe it. Unfortunately, the mathematics is too involved and we cannot go into details. Presented here is a polished and simplified version lacking many details and precise definitions. Nevertheless, for our purposes, details are not that important and the "big picture" will do fine.

By the time Heisenberg entered the stage, quantum theory was already in full swing. Planck had introduced the quanta, Einstein had published the "Photon paper," Bohr had introduced his model of an atom. One of the biggest challenges at that moment was related to the intensity of spectra. In a nutshell we have the following: Spectral lines of light had been studied extensively and the theory behind them was well understood. The whole idea of spectra refers to waves and, as long as we treat light as a wave, things work fine. However, the quantum assumptions coupled with Einstein's idea of a photon spoiled the picture, since a photon is a particle, not a wave. To make things worse, the mathematics used to describe the spectra was Fourier analysis, which is based on trigonometric functions that nicely communicate with waves but have nothing to do with particles.

The problem was a serious one and Heisenberg was one of many brave enough to tackle it. Soon though, very intense calculations bogged him down. He produced pages and pages of intricate mathematics, some of which did not follow full mathematical rigor. He switched from standard description of Newton's second law to Hamilton's version and made several "interesting" mathematical steps. He worked with matrices but did not know matrix algebra. Complex computation with matrices is often a hurdle even if one knows and applies sophisticated matrix algebra; and one can only imagine the horrific computations Heisenberg encountered without using the usual shortcuts.

Somehow, he managed to complete the work and derive the formula. And what a formula that was! This result is still one of the most wonderful tales related to quantum mechanics. It states that it is impossible to find out the exact *position* and *momentum* of a particle. In other words, if one knows the exact position of a particle one cannot know its velocity and vice versa. It is important to understand that this restriction is not due to the imperfection of our instruments but due to mathematics. Nobody, not even the creator, can compute these quantities simultaneously. This goes against intuition and against all the traditions in science where the experiment takes center stage. The ramifications of this astonishing discovery are enormous, but not really our cup of tea. Our quest deals with mathematics and we would like to know what role it played in Heisenberg's Uncertainty Principle.

And here it is. It boils down to the way we, mathematicians, multiply matrices. Namely, a matrix is nothing more than a table of numbers with n rows and m columns. For example, these are matrices:

$$A=\begin{bmatrix}2 & 3\\ 1 & 1\end{bmatrix} \text{ and } B=\begin{bmatrix}1 & 2\\ 0 & 3\end{bmatrix}.$$

Like most mathematical objects, these could be algebraically manipulated. That is, we can compute things like $A+B$, $5\cdot B$, $A\cdot B$, etc. Mathematical definition for some of these manipulations is intuitive. For example, $A+B$ is performed the way one would expect:

$$A+B=\begin{bmatrix}2 & 3\\ 1 & 1\end{bmatrix}+\begin{bmatrix}1 & 2\\ 0 & 3\end{bmatrix}=\begin{bmatrix}2+1 & 3+2\\ 1+0 & 1+3\end{bmatrix}=\begin{bmatrix}3 & 5\\ 1 & 4\end{bmatrix}.$$

The same is true for $7\cdot A$:

$$7\cdot A=7\cdot\begin{bmatrix}2 & 3\\ 1 & 1\end{bmatrix}=\begin{bmatrix}7\cdot 2 & 7\cdot 3\\ 7\cdot 1 & 7\cdot 1\end{bmatrix}=\begin{bmatrix}14 & 21\\ 7 & 7\end{bmatrix}.$$

While these two operations are intuitive, matrix multiplication is not. One would expect that multiplication works in the same way as summation. Something like this:

Incorrect matrix multiplication:

$$A\cdot B=\begin{bmatrix}a & b\\ c & d\end{bmatrix}\cdot\begin{bmatrix}x & y\\ w & z\end{bmatrix}=\begin{bmatrix}ab & by\\ cd & dz\end{bmatrix}.$$

But strangely enough, mathematicians did not define the multiplication in such a "natural" way. Instead, a cumbersome and counterintuitive definition is in place. See for yourself.

Correct matrix multiplication:

$$A \cdot B = \begin{bmatrix} a & b \\ c & d \end{bmatrix} \cdot \begin{bmatrix} x & y \\ w & z \end{bmatrix} = \begin{bmatrix} ax+bw & ay+bz \\ cx+dw & cy+dz \end{bmatrix}.$$

Unless the reader is "one of us," she probably wonders why on earth anybody would define multiplication in such an awkward way. Why would summation be part of matrix multiplication? But we had our reasons. I will not go in details, but suffice it to say that this way of multiplying matrices allows us to extend the idea of linear functions into multidimensional settings.

Sadly, one of the very important, and mathematically annoying, consequences of such a definition is the sacrifice of the commutative property. In contrast to numbers, for which $a \cdot b$ is always equal to $b \cdot a$, matrices do not commute. In other words, for matrices, $AB \neq BA$. Below is a contra-example:

$$A \cdot B = \begin{bmatrix} 2 & 3 \\ 1 & 1 \end{bmatrix} \cdot \begin{bmatrix} 1 & 2 \\ 0 & 3 \end{bmatrix} = \begin{bmatrix} 2+0 & 4+9 \\ 1+0 & 2+3 \end{bmatrix} = \begin{bmatrix} 2 & 13 \\ 1 & 5 \end{bmatrix},$$

while

$$B \cdot A = \begin{bmatrix} 1 & 2 \\ 0 & 3 \end{bmatrix} \cdot \begin{bmatrix} 2 & 3 \\ 1 & 1 \end{bmatrix} = \begin{bmatrix} 2+2 & 3+2 \\ 0+3 & 0+3 \end{bmatrix} = \begin{bmatrix} 4 & 5 \\ 3 & 3 \end{bmatrix}.$$

Thus it is clear that $AB \neq BA$.

Our poor Heisenberg did not seem to know this. He did not know matrix algebra, and, in all likelihood he did not even know that such mathematics existed (although it had been around for nearly a century prior to his time). But, genius as he was, he reinvented it, and among other things he multiplied the matrices in exactly same way as mathematicians did (for you see, nature finds it logical too). And now comes the grand finale, another bizarre mathematical event that shook the world of science. Namely, while "massaging" his matrices, Hermitian operators and such, while writing and rewriting different physical phenomena, and then inserting them into mathematical equations (as physicists often do), he stumbled onto a "curious" thing. (A simplified version of this computation can be found in the Appendix.)

In somewhat layman's terms, this is what happened: Solving a mathematical problem using one set of physical laws/interpretations, he got the answer in the form of AB. When he attacked the same problem, using a different approach, he got the answer BA. As we all know, the same problem must have the same solution, regardless of which path we take to solve it, and AB seems the same as BA. But it was not! In his case A and B were matrices, not numbers, and the results did not cancel. He expected that

$$AB - BA = 0,$$

but instead he got

$$AB - BA \neq 0.$$

And this is it! This is the very source of the famous *Uncertainty Principle*. Namely, in his case, the matrix difference (in essence) is the error ($|AB - BA| = |\sigma| > 0$), where σ^2 is the "standard error"; which cannot be zero. In other words, no matter how precisely we measure things, the error will not be zero. Why? Well, because matrix multiplication is not commutative. That is why we have the *Uncertainty Principle*. There are no "great mysteries," no "fundamental deep understanding." It all boils down to a simple, almost trivial fact that matrix multiplication does not commute.

3.4 Conclusion

The three stories here are my favorites. They are arguably some of the most fundamental discoveries of modern science, and yet each one is counterintuitive and unexpected. These discoveries were not related to experimental results, but rather mathematical ones. These laws of nature, which created new physics, are not the product of Galileo's school (i.e., experiments) but more in line with Pythagoras (pure reason).

Let us examine the role mathematics has played here. We will start with the most predictable of the three, Einstein's example. Mathematics' role was straightforward. Once the "wise man" produced his amazing postulates, the mathematical machinery was turned on and the stream of strange and crazy consequences (laws of physics/theorems) followed. So as mathematics goes, in this example, it was just a vehicle that allowed us to conclude the *inevitable*.

Planck's story also describes the interplay between mathematics and physics, but unlike Einstein's example it went the "other way around." Namely, it is mathematics that dictated this fundamental law of physics. The quantum postulate was not based on a great philosopher's intuition, nor was it experimentally observed. It appeared because mathematics demanded it. The only way to make peace between experimental results and mathematics was to introduce this new, strange law of physics: the quanta.

Heisenberg's story goes one step further than Planck's. It is an "other way around" example, for it was mathematics dictating the fundamental law of physics, but Heisenberg did this *without any experimental* results as a guide, nor did he have any intuition to back him up. It was just mathematics! Planck had black-body radiation as a confirmation while Einstein had his intuition, as well as the Michelson-Morley experiment confirming the second postulate. Heisenberg, on the other hand, trusted mathematics completely, even though some experiments pointed against his theory (the cloud chamber experiment "clearly" shows that electrons have measurable path and momentum).

This is all nice and interesting, but what about our task? Did we learn anything? Do we see light at the end of the tunnel? We turned the table and looked at this bizarre, prophetic power of mathematics, but now from a physics' point of view. We analyzed, quite carefully, some of the most important results of modern science, and, everywhere we turned, we found mathematical footprints. And yet, it seems, we learned nothing. Mathematics played the central role, of that we are certain, but why? We have no clue.

Intermezzo
What Have We Learned?

Let us pause for a second. Our guiding question has been: How is it possible that mathematical theories designed thousands of years ago, and for different purposes, play such a fundamental role in shaping the very laws of nature, laws that we have only recently discovered. We investigated this particular point in detail, and I am confident that the reader agrees: There is a question to be answered.

However, our attempts to answer this question, or even to shed some light on the underlying mechanism, seemed to have failed. We tried looking more carefully at the underlying mathematics, ancient and modern, with no success. We examined physics, and the way mathematics forced its way into the laws of nature, and still we do not seem much wiser. If anything, we are even more confused. All we did is to uncover a remarkable maze, a labyrinth, in which all roads seem to be dictated by the laws of mathematics. Like it or not, at every turn, we stumble on math.

But, as any scientist or mathematician knows, trying desperately, pushing and shoving, writing and rewriting math proofs, experimenting, trying different things, well, that is the way we do things. If a chemist or molecular biologist encounters a bizarre, unexplained biochemical reaction, she conducts numerous experiments under different circumstances, in different mediums, collects data, and tries to gain an insight, a "eureka moment." And if this does not happen, which is often the case, one must regroup, take a "time out," collect all the data, and try to make sense of it. And ask a question: Did I miss something? Or at least: What have I learned so far? Well, we have learned a few things as the following tale will illustrate.

Alien Board Game

Imagine a world where people have been given an *alien board game*; say a sophisticated game found in the wreckage of an alien spaceship. This game has elements of chess, Go, and bridge, but is much more complex. Imagine that people can play this game on some device, a computer maybe, and as a final twist, imagine that they do not know the rules nor do they have any instructions related to this game. Let us also assume that people who play the game are rewarded. The more points they get, the greater the reward. Humans would, naturally, assume that this alien game was logical (without contradictory rules), and that the essential rules were few and simple. Under this scenario, it is not hard to accept that after some initial setbacks, after lot of trial and error, the "players" of this alien game would finally manage to uncover some basic properties and a few fundamental rules.

But, in this imaginary world, let us assume, there also exists a strange *cult* that has been studying board games for millennia. Why they do this is not clear, but let us imagine that for centuries, long before the alien game appeared, these *monks* have been carefully describing all kind of board games. Games with two players, with three players, with n-players; games with or without random aspects, finite games and infinite games. They have characterized and described a vast collection of possible configurations, outcomes, and structures that a game can have. They have found that some games are simple and trivial (like "tic-tac-toe") and some are complex and interesting (like chess), some games are repetitive and some are not. As a bizarre twist, imagine that in this world the monks are not really interested in *playing* the games they study. They seldom, if ever, actually construct a board and play a game. All they do is study the structure of a game: any game. As I said, a strange cult.

It is pretty clear what is coming next. One should not be surprised to find that many of the fundamental rules and properties the players discovered while playing the alien game had already been described and characterized by the monks ages earlier. And why not? The monks had had a thousand year head start and plenty of time to successfully characterize a vast collection of different games. They understood the structures, configurations, and properties of so many different games, and it would be logical to expect that some of the structures in the alien game were prophesized by the monks.

I am sure the reader sees the analogy. The "players" are the scientists and the "monks" are the mathematicians. The scientists work backward. They practice "reverse engineering." They study the final product, the nature, and try to decipher the creator's blueprint. That is their "alien game." This is a painstaking process, but they work hard, and with a lot of trial and error, with some luck and some intuition they progress just fine. In the process of doing this, the scientists discovered one very important and undisputable fact: *The universe is mathematical.* It is describable in mathematical terms and usually the mathematical formulas involved are very simple and elegant. This turns out to be exceptionally helpful, for without this, the poor scientists would have had an immeasurable job to perform.

Mathematicians, on the other hand, began from the very opposite end. We do not build anything, of course. We play our game: Starting with a set of consistent axioms, any set, we develop our theories. Some theories turn out to be dead-end and some turn out to be very fruitful. The latter are further developed, enhanced, and typically connected to other fruitful theories. That is how the game called mathematics is played. Since we started doing this about 2000 years before Galileo founded modern science, we have developed a colossal collection of theorems, and it seems plausible that we predated many mathematical concepts that scientists encountered in nature.

I hope the reader agrees that there is something to this idea. If not "the solution," this narrative has to be a part of the solution. It all fits so perfectly. Nature is a machine that runs on its own. There is no question about that. The laws of nature are relatively simple and comprehensible. This, too, is undisputable. These rules are not contradictory, for otherwise the universe would have already collapsed. Thus, nature's rules look like one of mathematical concoctions: A small set of relatively simple, noncontradictory axioms. And we mathematicians have been studying exactly these types of situations: Mechanisms (theories) that arise from a set of a few simple noncontradictory rules.

However, as any self-respecting scientist knows well, we must be careful and self-critical. It is very easy to fool oneself and accept a theory that seems "too good to be a coincidence," only to realize, much later, that it is a coincidence. This scenario happens all too often. Two obvious, and fundamental, issues are still unanswered:

a. Even if we assume that nature is mathematical, why would we expect "A few simple mathematical rules"? Why not a few thousand very complicated mathematical rules?
b. If we assume that mathematicians are addressing *all the games,* then there are infinitely many different games (i.e., theories) to study, and our 2 000-year head-start does not seem like much. It is still extremely unlikely that the theories mathematicians describe are the ones nature uses.

4 On Computer Games

In this chapter we will turn the tables. We will be the creators of our cosmos; that is, we will create our own simulated universe, with our own rules and our own laws of nature. And very quickly we will realize how daunting a task we face.

4.1 Simulated Universe

The tale of the *alien game* in the Intermezzo suggests an intriguing course of action: Why not simulate our own universe? By doing so we could get a taste of the medicine; we could experience the difficulties the "great architect" faced when he created our world. We are free to create any universe and design any "cosmos" we want. This world can be as remote from the laws of physics as we wish. However, we must agree on the two basic criteria upon which this newly created universe will be judged:

1. *Consistency and self-sufficiency*: Our universe must run on its own and for as long as possible. An intervention from an outside intelligent being is strongly discouraged.
2. *Appeal and complexity:* Our universe must be intricate and interesting. A ball bouncing back and forth at a constant speed would be both self-sufficient and would last forever, but it would be a very repetitive and dull universe.

4.1.1 Some Examples

Imagine a computer program simulating Player A and Player B engaged in a game of tic-tac-toe. As soon as the game ends, the computer records the results, switches the order of players, and starts all over again. This would be a self-sustained universe that would run forever without any contradiction. Every single game would either end in a draw or a win. Of course, this would not be a very appealing universe. All the possible outcomes are known in advance and if we equip our "players" with some kind of artificial intelligence, a learning subroutine, the universe would quickly learn to play the best possible game and the game would always end up in a draw.

So far so good. Now let us try something more interesting: The setup is the same, but the "players" (in this case computer subroutines) are playing chess instead of tic-tac-toe. As before, in order to make the universe more appealing, one could enhance the players with learning subroutines. This would mean that in addition to the usual rules of chess, the "players" would obey several more rules and would have some kind of finite memory that would allow them to "learn" from mistakes. Depending on these additional learning rules we could vary the degree of sophistication our "chess universe" would exhibit. Nevertheless, even with the most trivial setup, a random chess game, this universe would still be much more exciting than the tic-tac-toe universe.

But how do we know that this universe would not encounter an infinite loop, an endless game, or a situation which could not be resolved without outside intervention. In other words: *How do we know if this chess universe is self-sufficient?* Programming the rules of a chess game is trivial and designing some learning procedures not too hard. Like tic-tac-toe, the game of chess could also end up with a *win* by one of the two players (checkmate) or with a *draw*. Thus, we are in a very similar situation as before. However, there is a problem, and it is not related to the checkmates (they are easily recognized by a computer). The problem is with the draw. Unlike tic-tac-toe, the game of chess does not offer an easy way to declare a draw. The basic game of chess, one we have learned as children, does not clearly stipulate it. In a game played between two humans, the vast majority of draws are declared by "mutual agreement." And you see the problem: "How to teach the computer, our chess universe, to agree?" *Mutual agreement* presupposes intelligence and emotions, and we stipulated a *self sufficient universe* without any intervention from an outside intelligent being. Why? Because that is how our universe runs. The laws of physics are the rules and our universe seems to be running without "outside intervention."

The International Chess Federation (FIDE) recognizes this problem and it offers five rules in order to clarify draw situations. However, a close inspection reveals that these "rules" are not precise, they heavily depend on human interpretation and they serve more as guidelines than a policy.

1. *Mutual Agreement*: This rule is impossible to implement without outside intervention.

2. *Stalemate*: If the player, on taking their turn, has no legal move but is not in check, this is stalemate and the game is a draw.

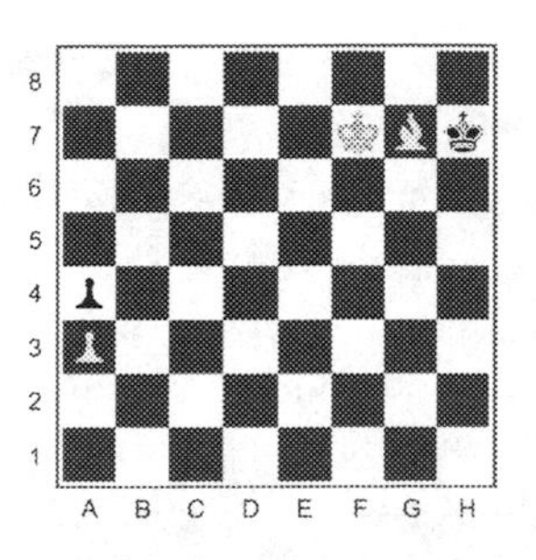

Figure 4.1 A stalemate.
Black is not in check but cannot move. This rule is easy to implement, and our "chess universe" should be equipped with it.

3. *The Impossibility of Checkmate*: If a position arises in which neither player could possibly give checkmate by a series of legal moves (because there is insufficient material left, as, for example, king and bishop against king), the game is a draw.

This is much easier said than done. Although in some situations one could indeed prove that a checkmate is impossible, there are countless other situations in which this decision is hard to make. There are plenty of known chess positions that would require an expert chess player (and dozens of moves) before checkmate is realized. So, we would need a grandmaster, a world-class expert in order to implement this rule. How else would we declare *The Impossibility of Checkmate*? In fact, even with a grandmaster at hand, one could still be in trouble. Maybe there are situations in which even the best of the players could not decide if a checkmate is possible or not.

I hope the reader realizes that we have a serious problem here. How would we tell the computer that Players A and B are in a draw situation? We could arbitrarily decide that if after certain number of moves a checkmate is not achieved, the game is a draw, but that would not be fair.

4. *Three fold Repetition*: If an identical position has occurred three times, or will occur after the player on turn moves, the player on turn may claim a draw.

There are two problems with this rule. First, it is not automatic. The player must claim it, which again presupposes that an intelligent, emotional being will judge the situation and decide on it. The second problem is more troublesome. Even if we make it automatic (by forcing a draw as soon as *Threefold Repetition* occurs), we still face a problem. It is very easy to construct a situation in which one has an endless chain of *twofold repetitions* that are interrupted with some simple moves elsewhere. In other words, one could design a complicated, cyclic, and repetitive motion that would run forever (and collapse our universe) but that would not violate the *Threefold Repetition* restriction.

Apparently, the International Chess Federation is aware of all the above defects and it realizes that the above rules do not prevent "an infinite game." For these reasons, the FIDE declared an additional rule that solves the problem (and sacrifices the essence of the game):

5. *Fifty-Move Rule:* If 50 moves have passed with no pawn being moved and no capture being made, a draw may be claimed (again, the draw is not automatic).

As our chess universe is not intelligent it cannot "agree." Thus, in order to implement this rule, one has to enforce it. With this enforcement we are guaranteed that no game can run indefinitely but we have substantially changed the game. The arbitrary nature of this rule is troubling and in many cases unfair (there are situations in which a player could have won, but it would have taken more than 50 moves).

Finally, even with all these rules, even after drastically changing the game, we are still not sure that our chess universe will not collapse. The 50-move rule prevents infinite games, but to the best of my knowledge there are no proofs, no mathematical guarantees, that some strange, unforeseen, stalemate-type situation could not happen. In other words, we still do not know if our chess universe is consistent.

4.1.2 The Tradeoff

The two examples above hint at a curious state of affairs. An interesting tradeoff has emerged:

Self-sufficient universe (tic-tac-toe) → *Dull repetitive universe*
Interesting universe (chess) → *Inconsistent universe*

Consistent games are typically boring while interesting games are typically inconsistent. It turns out that this curious tradeoff is not an accidental occurrence but a very systematic and fundamental mathematical property. The mathematics behind this phenomenon is very deep, and will be addressed shortly. However, before we do so, I would like to introduce yet another simple simulated universe. But this time the universe in question was designed by a mathematician.

4.2 John Conway's Game of Life

In 1970, Cambridge mathematician John Conway created an appealing little computer simulation, the celebrated "Game of Life." The rules are as follows:

On an infinite grid (an infinite chessboard) one places finitely many black squares, indicating that these cells are "alive," while the rest of the unoccupied (white) cells are declared "dead." The game follows these four simple rules (axioms):

1. Any live cell with fewer than two live neighbors dies (loneliness).
2. Any live cell with more than three live neighbors dies (overcrowding).
3. Any live cell with two or three live neighbors lives, unchanged, to the next generation.
4. Any dead cell with exactly three live neighbors comes to life.

Here, by "neighbor," we mean the adjacent cells.

This universe runs on its own time. That is, at each step all the cells are revisited, and the algorithm decides if a cell is black or white and then the computer prints out the result (a black or white cell). This universe, although it has a very simple setting and rules, is splendidly complex, rich, and interesting.

Let us investigate what happens with some simple initial configurations. Many of them either disappear or quickly collapse into static structures like "block" and "boat" in Figure 4.2 and 4.3:

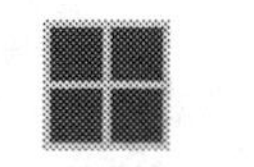

Figure 4.2 Block.

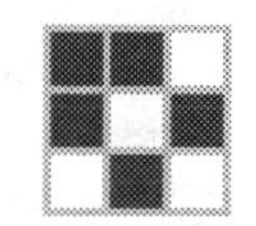

Figure 4.3 Boat.

They are static since every living (black) cell has exactly two or three living neighbors and according to Axiom 3 nothing changes.

Some configurations perform simple tasks. These "two-phase oscillators" repeat themselves after two steps. They are called "blinker" and "toad" (see Figure 4.4 and 4.5):

Figure 4.4 Blinker.

Figure 4.5 Toad.

However, a very simple modification can create unexpected lifelike behavior. For example, a starting configuration, very similar to the "boat" (the second static example Figure 4.3 creates a "glider" (see Figure 4.6):

Figure 4.6 Glider.

This is a special configuration. It does not collapse nor does it perform a "boring" two-step iteration. It does something much more interesting: It glides. Actually, its movement resembles wiggling, but that is not the point. The important thing is that a simple configuration exhibits interesting and unexpected behavior (for each iterative step it wiggles toward the upper-left corner).

Similarly, the "lightweight spaceship" travels in a zig-zag path toward the left (see Figure 4.7):

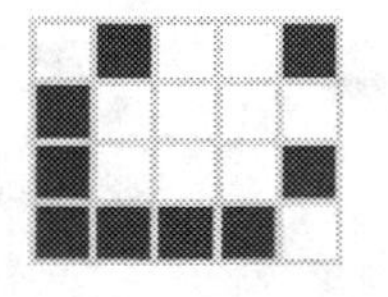

Figure 4.7 Lightweight spaceship.

Conway's Game of Life is extremely popular because there are countless interesting configurations that exhibit unexpected lifelike behavior. Take, for example, the "diehard" and "acorn" configurations shown in Figure 4.8 and 4.9:

Figure 4.8 Diehard.

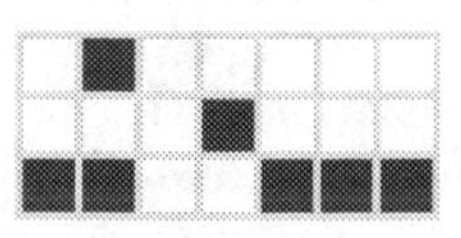

Figure 4.9 Acorn.

"Diehard" is a pattern that eventually disappears after 130 iterations while "acorn" takes 5,206 iterations and, in the process, it generates at least 25 gliders and oscillators. These few configurations, although interesting, are limited in one important aspect: They have limited growth. Conway's rules for birth are very restrictive: A new cell can be born only if it is surrounded with exactly three neighbors. Each of the above eight patterns will produce a universe whose number of living cells is kept from growing indefinitely. Initially, Conway was under the impression that his universe would forbid an infinite growth. More precisely, he conjectured that for any finite initial configuration, his universe would always have limited number of living cells (he even offered a cash prize for the solution).

Only a few months later, Bill Gosper produced the "Gosper Glider Gun" that exhibits infinite growth. The simplest configuration (with only 10 living cells) that grows indefinitely is the "10-cells" pattern shown in Figure 4.10:

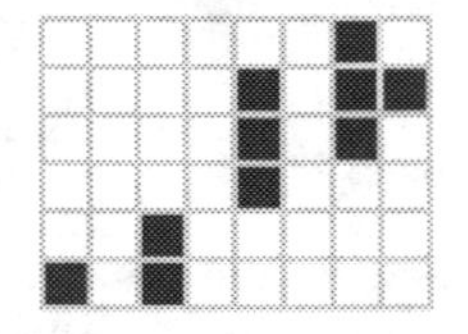

Figure 4.10 Smallest infinite growing pattern.

Other simple patterns that exhibit infinite growth are presented in Figure 4.11:

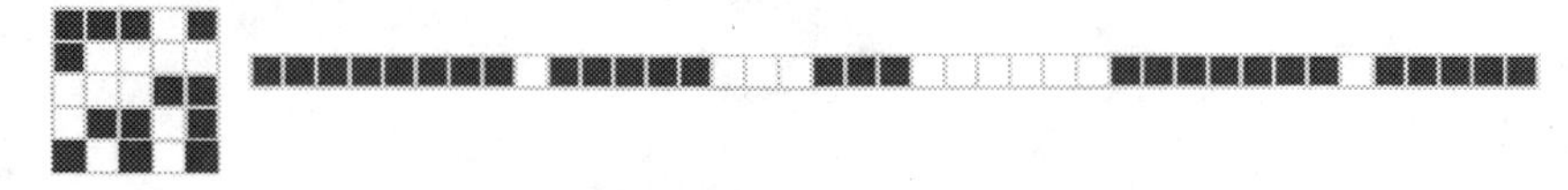

Figure 4.11 Infinite growing patterns.

The first has 13 living cells but it fits into a smaller square, while the second is one-dimensional. The above examples produce an infinite number of cells in a rather boring fashion, but the very first example, the "Gosper Glider Gun" (Figure 4.12) is actually more interesting. It produces (shoots out) its first glider on the 15th generation, and another glider every 30th generation from that point on. In other words, we have one very regulated mechanism that creates predictable patterns at predictable rate:

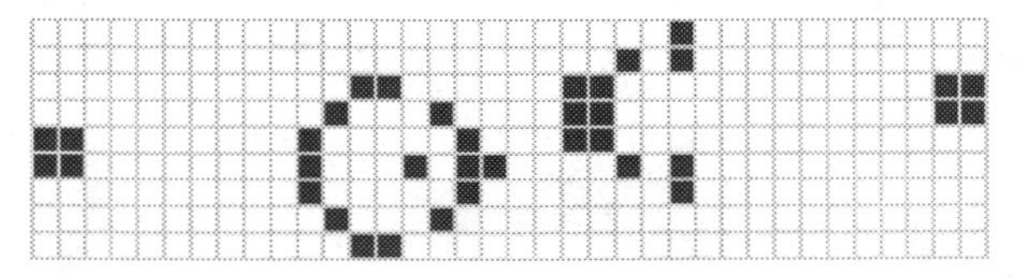

Figure 4.12 Gosper Glider Gun.

The literature is filled with numerous (thousands) of new configurations, some of them extremely elaborate, exhibiting many of the lifelike characteristics we see among microorganisms: They move, grow, divide, produce offspring, die. Many believe that one could actually create an initial pattern that would grow into a structure complex enough to be called "alive." It is easy to build upon the original design by adding new features: random components (death and birth is randomized), cells with different colors (adding to the complexity), limited cell memory (birth and death depend on current and previous states), finite grid (embedding the grid on closed-surface torus), etc.

Arguing whether Conway's Game of Life, or its modifications, could produce *bonafide* lifelike behavior is a tempting proposition, but also one that does not fit the framework of this book. On the other hand, discussing the origins and the properties of the initial rules (axioms) of the Game of Life does. First, one observes how nicely this universe fits between tic-tac-toe and chess. Its setting is almost identical to tic-tac-toe (a grid with black and white squares) and its axioms are extremely simple (in fact, it is easier to implement Conway's Game of Life on a computer than tic-tac-toe). Consequently, this universe is consistent and self-sustained. However, it is obviously not a dull universe.

So, it seems that this example contradicts my "tradeoff" assertion, for it appears that Conway hit the jackpot. The rules are complex enough to create an exciting evolving universe, but yet simple enough to prevent the consistency and self-sufficiency problems.

But this is exactly the point I want to make:

> In order to create such a universe, one really needs to hit the jackpot. One needs to perform an extremely difficult balancing act and create the axioms that can do both: provide a universe that is consistent as well as interesting.

It is not a coincidence that the creator of this game is a mathematician (and a world-class mathematician at that). As Conway himself testifies, it took him considerable time before he got it right. For example, Axiom 4 asserts that a cell can be born if it has exactly three living neighbors. But why three? If one changes this to any other assumption like "three or four living neighbors" or "exactly four living neighbors," or "two or four living neighbors," one gets a consistent but not a very exciting universe. I ran the simulations and, in each of these cases, very quickly (in matter of few hundred iterations), the universe collapses into a boring and trivial structure. Similarly, if one modifies any of the other three axioms, one almost always ends up with a consistent but dull universe.

The Game of Life is extremely popular and for decades thousands of programmers have been trying to create improved versions. Although the computers are a million times faster now, and although some interesting modifications of this game have surfaced, we have yet to see a true, fundamentally, and qualitatively better version of this game. The fact that we still do not have such a game is a testament, the final "proof," of how hard it is to create a simple, self-sustained, and yet interesting universe.

4.3 Bouncing Balls Universe

Now that we have learned about Conway's Game of Life, let us create our own self-sustained, and hopefully not completely trivial, universe. Our computer screen will be our "universe" and on it we will place a circle to act as a "ball." If we allocate some initial velocity to this ball and if we let it elastically bounce off the walls of our screen we get our primordial universe. This particular scenario is very easy to program on the computer. Plotting a ball on the screen is more or less trivial, and moving it in the appropriate direction is not much harder. All we need are two velocities, V_X and V_Y, and then, after a unit of time has passed, we erase the previous ball and plot a new one. But this time the center of the new ball shifts: V_X-units in the horizontal

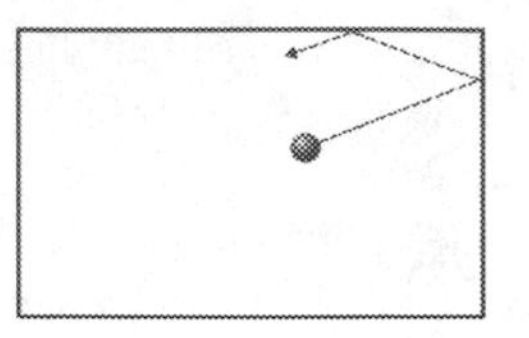

Figure 4.13 One-ball universe.

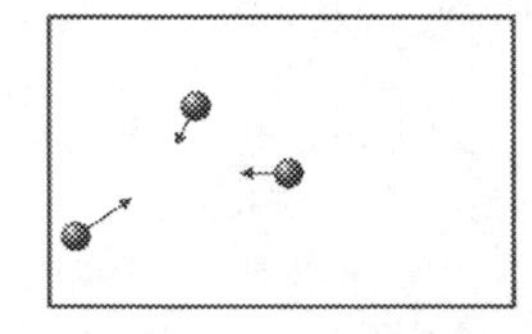

Figure 4.14 Multiball universe.

direction and V_Y-units in the vertical direction. This procedure creates the movement in the direction of the velocity vector $V = (V_X, V_Y)$. Finally, once the ball "hits the wall" as in Figure 4.13, one can trivially compute the new velocities ($V_X = -V_X$ for the impact on the vertical wall and $V_Y = -V_Y$ for the horizontal), and the "universe" continues like that indefinitely.

Thus, it seems, we have created a self-sustained universe, one than can run forever without human intervention. But, we must admit, this is a rather dull and repetitive universe. So, let us make it a little bit more interesting. Instead of one ball, why not put a few more in (k balls) and let them be of the same size, to make it simpler, but with different initial velocity vectors, to make it more interesting. And this universe, even with just three such balls, as in Figure 4.14, is quite an interesting one. It is fun to watch the collisions. Sometimes a ball will be "randomly" ejected in a strange direction and sometimes a ball will almost come to a dead stop, only to be hit a few moments later and hurled into an uncertain path. Quite interesting and (for some of us) almost hypnotic. If we add a few more balls (five to six works the best) then the fun never ends.

But there is a problem, and the fun does, indeed, end. To properly understand the issue we have to proceed slowly, one step at a time, particularly since what follows touches the very core of the ideas presented in this book. Namely, in the next few pages we will be both physicists as well as mathematicians. And, let us not forget, we are also the great architects of this simulated universe of ours. But, as they say, with great power comes great responsibility, so, in the lines that follow, the reader will "feel the pain" and hopefully appreciate how hard is to be the great architect. The very first problem we encounter is of a more technical nature, but it opens the door to another, rather strange issue, which then leads to a truly bizarre predicament.

But we are getting ahead of ourselves. Let us start with just two balls, and let us try modeling their collision. As we have seen, each ball (ball A and ball B) comes with velocity vector $V_A = (V_{A,X}, V_{A,Y})$ and $V_B = (V_{B,X}, V_{B,Y})$, where, as before, the quantities $V_{A,X}, V_{A,Y}$ stand for horizontal and vertical components. These are the known quantities before the collision. Now comes the fun part. We need to compute these vectors after the collision, and, in order to make our universe realistic, we will model this collision using the very laws of nature.

Since we simplified the problem (all balls have the same mass), we get the following equations:

(a) *Law of conservation of energy*:

$$(V_{A,X})^2+(V_{A,Y})^2+(V_{B,X})^2+(V_{B,Y})^2=(W_{A,X})^2+(W_{A,Y})^2+(W_{B,X})^2+(W_{B,Y})^2.$$

(b) *Law of conservation of momentum*:

$$V_{A,X}+V_{B,X}=W_{A,X}+W_{B,X}, \text{ and } V_{A,Y}+V_{B,Y}=W_{A,Y}+W_{B,Y}.$$

Here the W's stand for the velocities *after the collision*.

So far, this all makes sense. We have two balls, and since for each one we have two velocities (horizontal and vertical), we need four variables. Thus, in order to move the balls in our simulated universe, we need to compute these four unknowns ($W_{A,X}, W_{A,Y}, W_{B,X}, W_{B,Y}$). This brings us to the first (technical) issue. Namely, if we observe the above laws (the requirements a and b above) and the associated equations, we would count only three such equations. But we have four unknowns.

Luckily, there is a way out of this predicament. But it is a bit tricky. It involves angles and trigonometry. We need to put a new reference coordinate system (in the center of one of the balls), then we must account for the size of the balls as well as the angle of the collision. Not exactly very deep mathematics, but not trivial either. And it comes with a fairly cumbersome computation. Nevertheless, after all is said and done, we can recover another equation and we are good to go. We have four equations with four unknowns.

But, let us pause for a second here. What has just happened? Before we even blinked, this seemingly innocent, almost trivial little problem of two identical balls elastically colliding in a frictionless universe morphed into a mathematical problem of four (nonlinear) equations with four unknowns. Not exactly a walk in a park. Mathematics just pushed itself into our little universe, whether we wanted it there or not. And it gets worse.

Although we could easily implement the necessary algorithm and solve these equations, it is fair to say that in the process we have completely lost our intuition, as well as control over "our universe." For example, if we observe the two billiard balls in Figure 4.15 (both moving) that are about to collide, many of us would feel as if we have a good intuition about what is going to happen. We could "see" where the balls will hit each other and we have a good idea of how they will bounce off each other. Well, our intuition is wrong, for we have just learned that the balls will go wherever mathematics tells them to go, and we, frankly, have no idea where that might be. How many of us could instantly solve four equations with four unknowns?

The reader can easily convince herself of this fact. Namely, if one observes a billiard ball that is about to collide with a *stationary* ball, one has a very good intuition as to the balls' movements after the collision. This is the whole point of billiard games. On the other hand, if one tries to hit a *moving* ball in the game of billiards, one quickly learns that we have very little control (or idea) on the balls' movements after the collision.

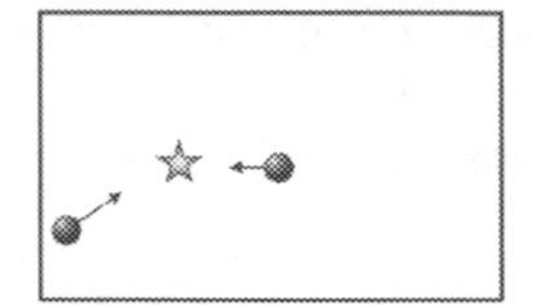

Figure 4.15 Two moving balls about to collide.

Yes, we might be the creators of this universe, but we have learned very quickly that it is mathematics that actually runs our universe and that we have very little control over it. The universe does what mathematics tells it to do.

This was the aforementioned "rather strange issue." The "star" of this chapter, the truly bizarre predicament mentioned earlier, is yet to come. And, to fully appreciate it, we need a small digression.

Many years ago, I actually implemented this particular program on my computer. I said to myself, instead of getting a commercially available screen saver, why not make one myself, with bouncing balls on the screen. The whole process took considerably more time than I intended, and I encountered numerous technical issues (and many self-induced computer bugs). Some of these issues I have already described, but some I have skipped (e.g., what to do if more than two balls simultaneously collide). Nevertheless, I succeeded, and it was worth it. With a click of a button, six balls would appear on the screen, each moving in a random direction and random velocity, and all one needed to do was to sit and relax, and spend (too many) hours observing this silly but fascinating little universe running all by itself.

Then the weekend came. Three days later, I walked back into my office, glanced at my custom-made screen saver and "hoot!" It was frozen. The six balls were on the screen, red balls on a bluish background, but they were not moving. "Another bug for certain," I thought. I had already spent too much time on this little project of mine, so I was just about to throw in the towel and get myself a "proper," commercially available screen saver. And then I saw it. Out of the corner of my eye, I saw one of the balls move. Just a tiny little bit, and just one of them, but move it did.

And this was significant. For you see, to write a computer program with a bug that would freeze the screen, well, that is rather common. But to have a bug that gradually slows down the screen, well that is an interesting bug. My universe was "leaking" energy somehow, somewhere. Namely, at the beginning, the universe was initialized with a fixed amount of kinetic energy, and as three days later the balls were barely moving, this energy had gone somewhere. This was particularly puzzling, since the very first equation I had implemented was specifically designed to preserve the energy. It forced the universe not to lose energy, and yet energy was lost.

To be sure, I double-checked this particular part of the code and indeed, the velocities after the collisions (the W's) and the velocities before the collisions (the V's) were correct. So, the bug was not there. I restarted the universe, but this time I printed (at the corner of the screen) the "total energy" of the system (adding the kinetic energy from all the balls), and there, in black and white, it stared at me: This number was gradually, almost infinitesimally slowly, going down. Tiny, minute amounts of energy were being lost. And at that moment it dawned on me. I knew what was going on.

It was the "round-off error." Namely, computers cannot hold all the digits of a number, but instead they keep only the first 8 (or 16) significant digits. For example, we know that $\frac{1}{3}$ = 0.33333333333..., with infinitely many decimals, but a computer does not know this. A computer, in its memory, keeps only 0.33333333, and consequently it "leaks" circa 0.000000003 amount of energy. In some cases, such as $\frac{2}{3}$ = 0.6666666666..., the computer will "round up" the last digit and keep 0.66666667. In this case, we have an increase of total energy.

At first glance, this does not seem like a problem. After each collision, depending on the invisible "ninth digit," our universe will either speed up a bit or slow down a bit, depending on how the computer rounded the last digit. Since it looks as if we are equally likely to add or to subtract this tiny amount of energy, everything should even out, right? But mathematics disagrees. What we get is a stochastic process, which inevitably collapses to zero. Mathematics is very clear about that: Even if we implement exactly 50% of the chances of adding or subtracting the energy, the total energy of the system converges to zero. The universe eventually collapses (see the Appendix). And there is nothing we can do. Except cheat.

Namely, the only way I could make my universe self-sufficient, so it could run forever, was to periodically compute the total energy of the system and then artificially add what was lost. This addition was very tiny and was evenly distributed among all the balls, thus to an uninformed observer invisible. But I hope the reader realizes the moral of this story: I had to cheat. For even this trivial universe, with two balls bouncing on the screen, requires human intervention. Periodically, it had to be "fixed."

Interestingly enough, on February 25, 1991, during the Gulf War, an American Patriot Missile battery in Dharan, Saudi Arabia, failed to track and intercept an incoming Iraqi Scud missile. The Scud struck an American Army barracks, killing 28 soldiers and injuring around 100 other people. It turns out that the cause was an inaccurate calculation due to the round-off error. The system's internal clock was multiplied by $\frac{1}{10}$ to produce the time in seconds, but $\frac{1}{10}$ in binary arithmetic has infinitely long representation and the software used only 24 digits (cutting off the rest). This rounding error accumulated over time. An easy calculation shows that after circa 100 hours, the resulting time error is about 0.34 seconds, in which time a Scud missile travels more than half a kilometer.

4.3.1 Since We Are on This Topic...

After the few lines above, one cannot help but ask the obvious question: How does our universe solve this "leaking energy problem"? Namely, if we accept the quantum postulate, and stipulate that smallest amount of energy is given by a fixed quantity (let us call it $\boldsymbol{q}$), then our real universe faces a problem not unlike the one we just described with our simulated universe. This is because, due to quantum mechanics, our real universe de facto operates with "finitely many digits," just like our computers. An example will help.

Imagine a situation in which two balls, each with energy of ½, collide in such a way that the first ball gets ⅓ of the initial energy while the second gets the remaining ⅔. It is easy to arrange a situation in which this would happen. However, since ⅓ has infinitely many digits, and since we cannot distribute the energy between the balls infinitely precisely, we have a problem. Namely, mathematics demands that we allocate ⅓ of $\boldsymbol{q}$ to the first ball and ⅔ of $\boldsymbol{q}$ to the second. But this violates postulates of quantum mechanics.

To be more precise: Mathematics (i.e., the system of equations) would produce the correct velocities $W_{A,X}, W_{A,Y}, W_{B,X}, W_{B,Y}$, but we (the universe) cannot use them. Why? Because these velocities would yield that ball A gets ⅓ of the energy, which is impossible, for this would "split the quantum." One way around would be to "cheat." Namely, one could modify the resultant velocities slightly by adding or subtracting a minute amount ε, in such a way as to satisfy both conditions: *to preserve the energy of the system and to satisfy the quantum postulate*. And although it is possible to perturb the required velocities in such a way, this perturbation would now distort the second postulate, one that preserves the momentum.

For better or worse, we have just demonstrated that the aforementioned three postulates – *the quantum postulate, preservation of energy postulate,* and *preservation of momentum postulate* – are contradictory. It is very easy to construct a situation in which one of the postulates must be violated. I am not sure how modern quantum mechanics resolves this situation, but, to be frank, that is not my "cup of tea" (see the Appendix). Our goal for this chapter is to describe, in some detail, how hard it is to be the great architect – how difficult it is to create a nontrivial game that would run on its own, without the creator's interventions.

4.4 Conclusion

Our universe, one with galaxies, atoms, and DNA, is self-sufficient and interesting. Its axioms, the laws of physics that is, are remarkably simple and comprehensible, and yet the final result, our world, is astonishingly complex. So, it seems that our "creator," the great architect has got it right. The "game" he created is not dull, yet it seems consistent, without contradictions.

And by all accounts our world is nothing but a game. Maybe even a simulation, for who could tell otherwise? Imagine if we could actually create a universe based not on an extension of Conway's Game of Life, but on a grid that is a trillion times bigger. Imagine if a life would appear there. What would be our role? Creators? Yes, but would we really be in control? Recall our trivial universe with two bouncing balls and how quickly our role changed *from a creator to a bystander*. How quickly mathematics took over. And one wonders: What if the creator of our universe had become a bystander, a servant to mighty oracle of mathematics?

5 On Mathematical Logic

We have tackled our issue from three different angles, mathematics, physics, and computer games, and each time we have stumbled onto the same topic: the axioms. We might call them postulates or rules of a game, depending on the angle from which we tackle the problem. We have also glimpsed a few laws that seem to govern this world of axioms. They tend to be very simple and come in small batches. And even then, even with only a few logical axioms, we often get a contradictory universe (or a game), like we did with chess and bouncing balls. Thus, the time has come to study the axioms themselves. Which brings us to this chapter on mathematical logic.

5.1 The Issue of Consistency

I hope the reader is convinced that in many ways mathematicians are also playing this "create your own universe" game. Countless years before computers existed and before "in silico" simulations were available, we were playing this game. We too start with a few basic rules (we call them axioms), and then with the help of logic and pure reason, we deduce the properties that follow. We are allowed to add new definitions, and to study new objects, but we are seldom allowed to change the rules.

Players of chess or the Game of Life investigate positions and new constructions (e.g., Queen's gambit and Gosper Glider Gun) while mathematicians study theorems, lemmas, and propositions. We do not use computer simulations to "uncover" the interesting positions, but rather pencil and paper. Nevertheless, the big picture is the same: We also create a universe, and this universe is judged on the very same criteria introduced earlier, *consistency* and *appeal.*

Appeal has always been the driving force behind mathematical research. For millennia, theorems have been judged with respect to their importance, beauty, and appeal. The reader, unless a mathematician, might be skeptical, for one often associates mathematical formula with adjectives like "dry" and "cold." Nevertheless, when a certain threshold in mathematical training is reached, the beauty becomes apparent, and one can appreciate an elegant proof and a beautiful lemma. As Hardy put it: "The mathematician's *patterns, like the painter's or the poet's must be beautiful; the ideas like the colors or the words, must fit together in a harmonious ways"* (*A Mathematician Apology,* Cambridge University Press, 1967, p. 84).

The first criterion, the issue of *consistency*, entered the picture quite late. Maybe too late. As we have seen with the chess example, it is quite possible, even expected, that perfectly logical, seemingly flawless initial rules can create a universe that is not consistent. Why would mathematical rules be any different? The game of mathematics is played with the help of logic and pure reason and one might wonder how we could have inconsistent mathematics. All we need is to derive "perfectly logical" arguments from the initial axioms. The predicament, of course, is *how do we know that the initial set of axioms is consistent?*

Remember the stalemate and chess example in Chapter 4? It took a thousand years before the first stalemate was officially described (1422 Cracow manuscript), and for a while, different regions had different ways of resolving it (in England, for centuries, it was considered a loss if a player's king was in a stalemate). There are no guarantees that mathematical axioms (rules) would not exhibit the very same contradictory behavior. Who is to guarantee that thousands of years from now, someone could not produce a stalemate-like situation in algebra or number theory?

An example will help. Consider the following three axioms on "children and people":

Axiom 1. All people are smart.
Axiom 2. All children are people.
Axiom 3. There exists a child who is not smart.

Clearly one can see that Axiom 3 is in contradiction with Axioms 1 and 2. The contradiction is not direct, since no individual axiom is in direct contradiction with any other individual axiom. To prove the inconsistency, one needs to perform an additional step: namely Axiom 1 and Axiom 2 yield Not Axiom 3 ($(A_1 \wedge A_2) \Rightarrow \neg A_3$).

This is an example of a simple and yet inconsistent universe. But what if the inconsistency is not so obvious? What if after a few centuries someone uncovers an inconsistency (mathematical stalemate) and proves that $2 + 2 = 5$, without violating any rules or axioms? Impossible? But how do you know? This might come as a surprise to you, but throughout history we have had several "close calls."

5.2 The Fifth Axiom (Again)

We have mentioned this example a few times already, but it pays to revisit it. The setup is as follows: Since antiquity, Euclid's Five Axioms of Geometry have been considered the cornerstones of mathematics. They serve as the exemplar, a blueprint if you will, of *how mathematical theory should develop*. The first four axioms are clear as day and one could hardly offer any argument against them.

5.2.1 Euclid's Axioms of Geometry

Axiom 1. It is possible to draw a straight line from any point to any point.
Axiom 2. It is possible to produce a finite straight line continuously in a straight line.
Axiom 3. It is possible to describe a circle with any center and distance.
Axiom 4. All right angles are equal to one another.
Axiom 5. If a straight line falling on two straight lines makes the interior angle on the same side less than two right angles, the two straight lines, if produced indefinitely, must meet.

The fifth axiom, on the other hand, is a bit tricky. It seems as if we could derive it from the first four, or if not, then maybe we could replace it with something more elegant, like the one-line statement of Axiom 2. This was attempted by many people over a long period, but without success. Nobody could eliminate the fifth axiom by deriving it from the first four, and all the attempts to replace the fifth axiom yielded statements that were of the same type: a bit clumsy, too descriptive, and not nearly as elegant as the first four axioms. The most common version of the fifth axiom is:

Axiom 5. (Alternative) Given a point P outside a line L, it is possible to draw only one line that is parallel to the line L and that meets point P.

Enter Lobachevsky. His reasoning was in many ways similar to our example above and the axioms on "children and people." Remember, Axiom 3 there was obviously in contradiction with the rest. Lobachevsky asked: What if we make the "fake" fifth axiom? One that should clearly lead to a contradiction.[1] His fifth axiom was (Figure 5.1):

Axiom 5. (Lobachevsky) Given a point P outside a line L, it is possible to draw more than one line that is parallel to the line L and that meets point P.[2]

Common sense clearly tells us that this axiom cannot possibly be right. Thus, all Lobachevsky needed to do was to use logic and pure reason, and derive a contradiction (as we did in our earlier example with "children and people" axioms). This, in turn, would demonstrate that we need the fifth axiom, since using its negation would yield an inconsistent theory. But then a funny thing happened. Lobachevsky was proving theorems using this bizarre axiom, and while these theorems clearly ontradicted common sense (angles inside triangles did not add up to 180°; parallel lines could asymptotically intercept, etc.) they were not logically inconsistent. Lobachevsky was a mathematician, not a physicist, so the fact that his results went against our feeble "common sense" did not matter. For him the only guide was logic and pure reason, and from that point of view there were no problems. These

[1] This approach was first tried by Saccheri and it is not exactly clear if Lobachevsky's train of thought was the same. The above narrative is based on private communications (from fellow Eastern European logicians) and it is included mainly for the sake of storytelling.

[2] Images are from Lobachevsky's original paper of 1829 "On Concise Outline of the Foundations of Geometry." Interestingly, the paper was first rejected by the St. Petersburg Academy of Sciences.

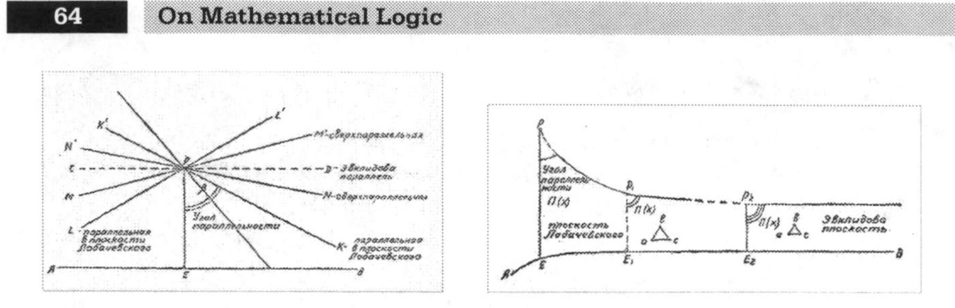

Figure 5.1 Lobachevsky's charts.

ridiculous, completely "useless" theorems were all logically correct and seemingly consistent. This, of course does not mean that there are no contradictions; it only meant that Lobachevsky could not find any.

Now do you see the problem? Although Lobachevsky, and later on Bolyai, Gauss, and numerous others, produced hundreds of theorems, propositions, and lemmas using these strange axioms, and although all of them appeared to be in perfect logical harmony, without any contradiction, this does not mean that a contradiction does not exist. It only means that one has not been found yet. Maybe after thousands, or even ten thousands of theorems, one could stumble on a contradiction.

As a final insult to injury, Italian mathematician Eugenio Beltrami proved that Lobachevsky's geometry admits an interpretation in terms of Euclidean geometry. From here it follows that if Lobachevsky's geometry leads to a contradiction, Euclidean geometry is contradictory too. In other words, the consistency of Euclidean geometry implies the consistency of the geometry of Lobachevsky. And yet, we still do not have proof that either geometry is consistent.

5.3 L'Axiome du Choix (The Axiom of Choice)

In slightly simplified version, the Axiom of Choice states the following:

> *If one has a collection of non empty sets, one can pick an element from each of these sets and combine these elements into another set.*

For example, suppose we have a set of *all people*, a set of *all fishes*, and finally a set of *all books*. Then, Axiom of Choice allows us to pick one person, one fish, and one book, and these three objects, "a person," "a fish," and "a book," if combined will form another set. It is as simple as that. This seems so trivial and obvious that the reader must be puzzled: You need an axiom for this?[3]

Well, it turns out that this "innocent" little axiom kept popping up in some very unexpected and controversial places. Some of the weirdest mathematical results are the direct product of this axiom. Arguably, the most controversial of them is the "doubling of the sphere" theorem (also known as *Banach-Tarski Theorem*). In slightly simplified form it states:

[3] Actually not. For the sake of simplicity, this example is based on finite number of sets, and the Axiom of Choice is not needed. Formally, Zermelo-Fraenkel Axioms would suffice.

Given a sphere of radius 1, one can break this sphere into finitely many pieces, take these pieces and reassemble them (without any deformation) into a sphere of radius 2.

This is a truly bizarre result. It clearly contradicts common sense as well as several laws of physics. Imagine Christmas decorations and those ornamental balls. Now imagine breaking one of them into finitely many pieces and reassembling these pieces into a ball twice as big. Actually, one can do this with only five pieces. The proof of this result fundamentally depends on the Axiom of Choice. By using this axiom, we "bypass" the issue of how to construct such a doubling, or how to break the sphere. We only know that such a doubling exists. How to actually perform this task is wrapped in the mystery of the Axiom of Choice.

The reader might be under impression that we are in a situation very similar to one described earlier with the "children and people" axioms. In that case, Axiom 3: "There exists a child that is not smart," created a contradictory statement and one had to discard or change that axiom. Thus, it seems, we should dismiss the Axiom of Choice for it has yielded a contradiction. But has it?

The Axiom of Choice certainly contradicted common sense and the laws of physics but it did not violate any law of mathematics. What about the area, you might ask? Since we did not deform the pieces of the sphere (just rearranged them) we did not change the area. However, the surface area of a sphere with radius r is given by the formula $4r^2\pi$, and since in one case $r = 1$ and in the other $r = 2$, it seems that the Banach–Tarski Theorem implies that $4\pi = 16\pi$, which yields $1 = 4$. And this ought to be a contradiction.

It turns out that there is a logical flaw in the above argument (which will be addressed shortly) and, consequently, the "doubling the sphere theorem" does not contradict any mathematical result, although it comes close. So close that some mathematicians reject this axiom. (Yes, mathematics, like any other religion, has its own orthodox faction. This group follows constructivism. In "essence," they believe that in order to "pick" an element from a set one needs to provide a construction, an algorithm that can select the particular element. One cannot just use the Axiom of Choice and declare: "Pick an element.") And this brings us to our next story.

5.4 The Problem with Measure Theory

Some two thousand years after Euclid's Axioms of Geometry and about one thousand years after many geometric figures and solids had been carefully studied, mathematicians realized that one of the most basic and most intuitive attributes related to geometric figures had not been carefully addressed. This aspect was *area*. One only needs to pause for a second to realize that there is something awry with the very idea of *area*.

Sure, the area of a square with side a is declared to be a^2. Everybody knows that. Similarly, the area of a rectangle is defined as a product of its sides. But what about other figures? We have a formula for the area of a circle, and an ellipse, and many other regular shapes, but that is not what I am asking. No, my question is actually much simpler: I am only interested in the definition of *area*. What does one think when one says, *the area of a figure?* How would one even define the area of some bizarrely shaped figure?

This turns out to be a rather tricky question. Mathematicians addressed it in their usual way: We set up a collection of axioms that would extend the definition of "*area of a square*" into the area of any subset of a plane $A \subset R^2$ (A "subset" $\subset$ of a "plane" R^2).

In other words, we set up a function λ, which is defined for any subset A, such that $\lambda(\text{Square}) = \lambda([0,a] \times [0,a]) = a^2$. Next, in Euclid's fashion, we listed the few essential, fundamental properties that this function must possess if it is to be called *an area of a set*. A slightly simplified version of these axioms, and in a more familiar setting, is presented below.

Imagine a kitchen table (of any size and shape) and try to "describe" its area. There are many properties the area (this function λ) should have, but the most essential, the most fundamental, properties are as follows:

Axiom 0: If the table is a square then its area is a^2.

However, if the table has a more complicated and irregular shape, then we need to impose a few additional rules:

Axiom 1. The table area cannot be negative (that is, $\lambda(A) \geq 0$*).*
Axiom 2. If you take a subsection of the table, its area cannot be larger than the area of the whole table (that is, $A \subset B \Rightarrow \lambda(A) \leq \lambda(B)$*).*
Axiom 3. If you move the table around the kitchen, or if you flip it over, you cannot change its area (that is, $\lambda(A) = \lambda(-A)$*) and* $\lambda(A+a) = \lambda(A)$*).*
Axiom 4. If you break the table into many parts the areas of each of the individual parts add up to the whole area of the table
(simplified: $A \cap B = \{\} \Rightarrow \lambda(A \cup B) = \lambda(A) + \lambda(B)$).

I am sure that the reader does not object to any of these axioms. They constitute the bare minimum an "area" should possess and they seem even more intuitive than Euclid's Axioms of Geometry. With this new function (that is, $\lambda(A)$) we could build the theory and define the area of any shape (any subset of the plane), no matter how crooked or irregular it might seem. We are ready to develop *measure theory.*

But there is a problem. These seemingly trivial, perfectly logical assumptions, these Axioms of Measure, are in contradiction. It is possible to prove that the function λ *does not exist.* In other words, it is *impossible* to define "area" for all the sets. Strange, isn't it?

If the reader is not a mathematician, she might object: *What do you mean, that* λ *does not exist?* Didn't you just define $\lambda(A)$ as the area of set A? That is, $\lambda(\mathit{Square}) =$ *area of a square* and $\lambda(\mathit{Circle}) =$ *area of a circle.* Right?

True, for some simple sets A, one could indeed define the function $\lambda(A)$, but what we wanted is a function that would work for *all the sets* A (for all tables, no matter how crooked). And this we could not do. The Axioms of Measure, as stated, are contradictory; they are not consistent. It is possible to prove that Axiom 1, Axiom 3, and Axiom 4 imply the opposite of Axiom 2. In other words, one can construct a subsection of a table that would have an "area" that is greater than the "area" of the whole table. This is an obvious contradiction. And it looks very familiar. Our silly example about "children and people" in Section 5.1 yielded a similar situation. There we showed that Axiom 1 and Axiom 2 implied Not Axiom 3.

For the details of the proof, the reader can check the Appendix, but I warn you: The proof is beautiful, short, logically perfect, and consistent (as long as one accepts the Axiom of Choice), but if you are expecting to see an intuitive and constructive proof, you will be disappointed. The construction of the contradictory set relies heavily on the Axiom of Choice, which states that we can "pick" desired elements, but it does not specify how to do so. Consequently, in this proof, any intuition or common sense is lost and what remains is just "cold" and "dispassionate" logic. Or, as some would say, beautiful and elegant logical argument.

As a consequence, mathematicians needed to revamp their measure theory. The above inconsistency implies that some sets are not measurable (since they lead to a contradiction) and consequently modern measure theory treats only the "measurable" sets. How to describe them and how to work with them will not be addressed here, but all we need to know is that the aforementioned *Banach-Tarski Theorem* (the doubling the sphere paradox) involves finitely many subsets of a sphere, and although these subsets are perfectly well defined as sets, they are not measurable sets. They are sets (pieces of a sphere) that do not have area. Their area is not zero. No, they just do not have any area.

Confused? You should be. Modern mathematics could only treat these "measurable" sets and the rest had to be left alone. It is a price we had to pay to satisfy mathematical logic and we learned to live with it. Therefore, the above argument that seemingly leads to a contradiction "$1 = 4$," is not logically correct. Let us revisit the mentioned procedure.

We used the fact that a sphere of radius r, S_r has area given by the formula $A_{S_r} = 4r^2\pi$. Next, by the "doubling" theorem there exist finitely many disjoint pieces (actually five pieces), such that $S_1 = P_1 + P_2 + P_3 + P_4 + P_5 = S_2$ (here by "+" we mean the union of disjoint pieces).

Therefore, one is tempted to conclude:

$$Area(S_1) = Area(P_1 + P_2 + P_3 + P_4 + P_5) = Area(P_1) + \dots + Area(P_5) = Area(S_2),$$

and from here the statement "$1 = 4$" follows. Except that $Area(P_1)$does not exist. The sets P_1, P_2, do not have an area, they are non measurable and formally (de jure) we did not make any logical mistake. But I hope the reader realizes how close to a contradiction we came.

5.4.1 *Final Thoughts*

The example of measure theory touches the very essence of this book. Namely, we (mathematicians) started with a set of rules (axioms) that were as clear as day, rules that were as intuitive as the Euclid Axioms, and yet they yielded a contradictory theory. Which raises two points:

1. It seems very hard to create a universe, or a game, or a mathematical theory that is consistent.
2. How do we know if our mathematical theory is consistent?

The built-in contradiction of measure theory is far from obvious and it took some very sophisticated mathematics and ingenious argument to uncover it. Luckily, we discovered this inconsistency very early on, for one could only imagine the consequences had we proceeded for another 1 000 years (as with chess and stalemate). I hope the reader can see the danger now. What if some other theory, like set theory or number theory, has a built-in deficiency, but what if this deficiency is so subtle that we have not detected it yet? What if someone can indeed prove that *1 = 4* and consequently destroy the number theory? And this brings us to yet another tale.

5.5 Russell's Paradox

For most of us mortals, the natural numbers, the integers 1, 2, 3, 4, and so on would be considered as the very basic blocks of mathematics. They represent something so basic, so essential that one cannot imagine any further simplification. This notion prevails through most of the civilized world and world of mathematics as well. It was Kronecker who said:

> God gave us the integers, all else is the work of man.[4]

But integers are not as simple as they seem. There is plenty of historical as well as psychological evidence that casts a doubt on the above claim. Let us start with children. If one observes children's development carefully, one realizes that these learning machines, these little "sponges" that absorb incredible amounts of information, rules, and logic, do not do well with numbers. They perform exceedingly complex face recognition analysis effortlessly and acquire complex rules of grammar with ease but with comparable trivial basic arithmetic or even just counting they struggle. One has to repeat the mantra *one-two-three-four-...* many, many times before a child accepts it. Not only does it take forever for them to hardwire the mantra *one-two-three*, but even when they accept these words, children do not really understand the idea of counting.

Consider the following experiment, which I performed with several extremely intelligent three-and four-year-old children. Put a few candies on a table (say four candies), select a child that just recently acquired the knowledge of numbers (the *one-two-three* mantra), and ask her to count the candies. First, you observe that most of the children have no idea what you mean when you say *count them*. Consequently, you need to help her by saying *one* (pick one candy), *two* (pick another candy), and so on.

What one observes next is quite surprising. The majority of children will just copy your procedure, they will pick a candy and say *one*, pick another candy and say *two*, and so on. But they often do not stop at four (as expected since there are four candies). Most of them just continue with *five-six-seven....* The idea that there is some kind of magical connection between the mantra (*one-two-three*) and the actual objects is not a simple thing to acquire. It takes several more months, sometimes years, and countless trials with different objects (toys, fingers, people, etc.) before children accept this seemingly obvious and fundamental point.

[4] Eric Temple Bell, *Men of Mathematics* (New York: Simon and Schuster, 1986), p. 477.

In fact, it is only because modern civilization puts such emphasis on numbers and mathematics that most parents and preschool teachers spend countless hours gently brainwashing children into accepting numbers and counting. Since the vast majority of us were brainwashed in this way, we now perceive the idea of a number and counting as so fundamental and so simple that one should not even bother trying to understand it any further. Why should we? Most of us learned it and accepted it before we even saw a farm animal, or snow, or an ocean. This idea was imprinted on our brains by our parents even before most of the grammar rules of our mother tongue.

But it was not always like this. As any anthropologist will testify, the majority of early humans (and today's primitive tribes) did not bother much with numbers. Essentially, all primitive cultures have words for *one* and *two* and seldom any more. In some places they developed numbers larger than two, but they would use different numbers for counting different things, which actually makes sense. So the word "five" for *five* kings is different from the word for *five* people and different again from *five* dogs. The idea of unifying all the objects into just "five" took a long time to develop.

This strongly suggests that more fundamental, more intuitive objects than numbers should be chosen as building blocks. It turns out that such an object, a logical construction, exists and it is called a *set*. A set is deemed so fundamental that one does not bother defining it. Georg Cantor, the principal creator of set theory, gave the following "definition" of a set:

> By a set we understand any collection M of definite, distinct objects m of our perception or of our thought (which will be called the elements of M) into a whole.[5]

This is a rather convoluted definition, and for a reason. The very idea of a set is so primordial that it is very hard, if not impossible, to define it using any simpler construction. (Cantor's definition is circular, since he describes "a set" by "a collection".) In reality what one does is to define a set by describing its elements.

In simple examples, we can list all the elements:

$$\text{Set } A = \{1,3,4\},$$

$$\text{Set } B = \{\text{Bob, John, } 35, a, b, c\}.$$

Some sets have infinitely many elements, but we still could describe them as a "list":

$$\text{Set } C = \{1, 3, 5, 7, 9, \dots\},$$

or a verbal description:

$$\text{Set } C = \{\text{all odd numbers}\}.$$

But not all sets are such that one can actually list the elements. One can just describe the property the elements should have:

$$\text{Set } D = \left\{\begin{matrix}\text{All people who have exactly 7 brothers,} \\ \text{have blue eyes, and are named Hassan}\end{matrix}\right\}.$$

[5] These are the first lines of Cantor's final paper: "Beiträge zur Begründung der transfiniten Mengenlehre," published in *Mathematische Annalen, vol.* 46 (1895): 481–512.

Although D is definitely not an infinite set, it is very likely that nobody on this planet could precisely list all its elements. Moreover, we are not even sure if this is an empty set or not. But that does not matter. We know that D is a set. What matters is that these objects, "the sets," are intuitive, easy to manipulate, and form a natural starting point for any mathematical theory. And it is not only the mathematics that is appealing. Unlike numbers, sets are very deeply embedded in the human psyche.

Little children intuitively know a set when they see one. The set of all adults is different from the set of all children. Sets are easily manipulated on some very primitive level: Combining the sets of Christmas presents from parents and grandparents is a *union* of two sets and no kid in the world needs hours of practice with the mantric *one-two-three* to understand that. Similarly, the candies that children can have after dinner are just a *subset* of all the candies in a house. The axioms and ideas behind set theory are far easier to understand and teach than counting, and one should perhaps correct Kronecker's quote to: *God gave us sets, all else is the work of man.*

Using sets, Bertrand Russell was able to define a number. In my opinion, this definition is one of the prettiest logical constructions. I cannot help but present it here:

> The number of a set is the set of all those sets that are similar to it.
> A number is anything, which is the number of some set.[6]

Such a definition has the verbal appearance of being circular, but in fact it is not. It is mathematically perfect and it also makes sense (if one tries it out a bit). Allow me to elaborate. A person walks into your room and says: *five*. What comes to your mind? Five what? Is it five people, five monkeys, five cakes? You have no idea. At that moment, number *five* represents *all of them*, all the sets that have the same number of elements (are similar to each other). Thus, number five is a collection of all these sets simultaneously. That is the essence of Russell's definition.

The definition is not circular since one can determine if two sets are *similar* (i.e., have the same number of elements) even *without knowing how to count*. In fact, most primitive tribes and young children can quickly determine if two sets have different cardinality (different numbers of elements) without ever using the idea of a *number*. This is very easy to confirm: Just give two siblings a different number of candies.

So how do they do this? An example will help: Even today some shepherds in remote areas "count" their sheep using the following trick. They have a collection of little sticks, which they keep in their pocket. Each stick corresponds to one sheep. At the end of a day, when they herd the flock, they just assign one stick to one sheep and move a stick from one pocket to the other. If all the sticks are removed from the pocket, all the sheep are accounted for. If there is a stick left, a sheep is missing. No need to count, no need for the *one-two-three* mantra.

That is the essence of Russell's definition. Namely, two sets are similar if it is possible to establish this one-to-one mapping (i.e., one-stick-one-sheep) and all the sets that are similar to each other will be called a *number*.

[6] The original definition appeared in Bertrand Russell's *Introduction to Mathematical Philosophy* (Macmillan, 1930). To keep the narrative consistent I changed Russell's *class* into *set*.

With this definition Russell (and Alfred Whitehead) were able to build integers, arithmetic, and essentially all of fundamental mathematics. But it was not easy: It took hundreds of pages before they were able to prove that 1+1=2. The reader, even if she is a mathematician, might dismiss these results as just a curiosity: *Fine, they prove that* 1+1=2, *but this is hardly an important question to ask.* Before we continue I would urge you to read, and read carefully, the following explanation offered by Russell himself:

> "But," you might say, "none of this shakes my belief that 2 and 2 are 4." You are quite right, except in marginal cases - and it is only in marginal cases that you are doubtful whether a certain animal is a dog or a certain length is less than a meter. Two must be two of something, and the proposition "2 and 2 are 4" is useless unless it can be applied. Two dogs and two dogs are certainly four dogs, but cases arise in which you are doubtful whether two of them are dogs. "Well, at any rate there are four animals," you may say. But there are microorganisms concerning which it is doubtful whether they are animals or plants. "Well, then living organisms," you say. But there are things of which it is doubtful whether they are living organisms or not. You will be driven into saying: "Two entities and two entities are four entities." *When you have told me what you mean by "entity," we will resume the argument.*[7]

And in many ways his results, although fundamental and important, would have stayed within the rather introverted clique of mathematical logicians. But fate had different plans for Bertrand. In the process of working on the fundamentals of mathematics he discovered a paradox, Russell's Paradox. In slightly simplified version it states:

> A set of all sets does not exist.

At first glance this neither looks like a paradox nor a particularly disturbing statement. So what if such a set does not exist? Well, let us think about it for a second. Remember our set above:

> D = {All the people who have exactly seven brothers, have blue eyes, and are named Hassan}.

We have no idea if this set has any elements at all. Even if D is not an empty set, we have no idea what its elements are. There are so many things we do not know about this set but one thing we know for sure: *It is a set.* What else could it be? The set D is constructed in a way that many important mathematical sets are constructed. We typically start with a general statement, like a set of *all people*, or a set of *all functions*, or a set of *all sets*, and then we list the desired properties, like *having seven brothers* (for humans) or *being continuous* (for function) or *being finite* (for sets). Many mathematical theorems begin with statements like these:

> Let H be a set of all functions that are continuous and bounded.
>
> Let K be a set of all sets that are subsets of a plane and are convex.

[7] From N. Rose, *Mathematical Maxims and Minims* (Raleigh NC: Rome Press Inc., 1988), p. 99.

Now you see the problem. Russell demonstrated that a "collection" of all sets is not a set. Thus, it is conceivable that our "collections" H and K are not sets either. If they are not sets, we are in trouble, for it is essentially impossible to derive any logical argument starting with something that is not a set. An obvious question presents itself: If these are not sets, what are they? Well, the only way to answer is: *I have no clue*. And, it is safe to say, nobody on this planet has a clue either. In the above statements I used the expression "collection," but this is just a semantic solution. In reality we cannot even comprehend a collection that is not a set. So how do we proceed you might ask? How do we prove that something is a set?

And here comes the final nail into the coffin: *We cannot!* The set is such a primordial structure that in general we do not have a mechanism that can guarantee that a certain constriction *is a set*. I hope the reader now appreciates the magnitude of this paradox. It shocked mathematics to its core and in many ways mathematics never fully recovered. It is a "lose-lose situation." There are constructions for which we could determine that they are *not sets* but in general we cannot prove that a construction *is a set*. Needless to say, Russell did not take this discovery well. Here is what he said:

> Mathematics may be defined as the subject in which we never know what we are talking about, nor whether what we are saying is true.[8]

But don't worry, Russell did just fine. He left mathematics (understandably) and devoted his life elsewhere. He became a prominent philosophical figure and activist (the Russell Tribunal) as well as writer (winning the Nobel Prize in Literature). Nevertheless, I strongly believe that a few hundred years from now, when everything is said and done and when the dust settles, his writing, his philosophy, and his activism will fade away. But Russell's Paradox is here to stay.

The proof of Russell's statement is actually not that hard. It is a masterpiece and I have to include it here:

> *Russell's Paradox*: Some sets are members of themselves, like the set of all sets and others are not, like the empty set. Let $A = \{x$ such that x is not in $x\}$. Is A in A? If so, it satisfies the property to be in A, therefore A is not in A. Is A not in A? Then it does not satisfy the property to be put into A, and therefore A is in A.

Perfect, isn't it? In many ways the essence of his paradox can be deduced from the following little puzzle:

> In village C, there exists a barber **a**, who shaves all the people that do not shave themselves (obviously). The question is: Who shaves the barber?

Clearly this was a shock and mathematicians needed to regroup and salvage the situation. The set theory needed a major reconstruction and this was provided by

[8] Rose, *Mathematical Maxims and Minims*, p. 97.

Zermelo–Fraenkel's Axioms. In essence, this approach resembles the measure theory "fix" described earlier: The setup had to be simplified and we have to limit ourselves to types of objects that could (should) be called sets. With this, Russell's and similar paradoxes are avoided, but, in my opinion, we paid a terrible price. Unlike the original, naïve set theory, this one cannot be taught to children. The original intuition, the elegance, and the beauty are lost. Unfortunately, this is the price we have to pay.[9]

5.6 Conclusion

The following story is so unbelievable that it must be true. After digesting these few examples – the Axiom of Choice, Russell's Paradox, and the Measure Theory Problem – the reader can easily foresee how mathematicians reacted. They were dazed and confused. The very pillars of mathematics were in jeopardy and the obvious questions presented themselves: Are we sure we have cleared all the mess? Maybe there are few other inconsistencies that we have not discovered yet? Can we prove that arithmetic after these fixes is still consistent, and if not, how can we fix it once and for all? What comes next dwarfs any Hollywood script or Shakespearean plot.

Enter Kurt Gödel. He accepted Hilbert's challenge, about arithmetic completeness. In other words, he needed to show that *every true mathematical statement is provable*. I am sure some readers will find his task confusing: Of course true statements are provable. How else would we know that they are true?

But if one thinks a little more carefully one indeed recognizes that it is entirely possible for some statements to be correct but for humans to never mathematically prove them.

Consider the following simple problem: Take any even integer number larger than two and check if it can be written as a sum of two prime numbers:

Take the number 4	4=2+2	Conclusion: Yes it can, since 2 is prime.
Take the number 6	6=3+3	Conclusion: Yes it can, since 3 is prime.
Take the number 8	8=3+5	Conclusion: Yes it can, 3 and 5 are primes.
	⋮	
Take the number 100	100=53+47	Conclusion: Yes it can, 53 and 47 are primes.

It seems that indeed *every* even number could be written in this form. This statement is the famous Goldbach conjecture and it has been confirmed for billions (actually trillions) of even numbers. Thus, there is a good chance that this statement is indeed correct. But can we prove it? For more than 250 years mathematicians tried and failed to prove this fact. I am sure that many a reader believes that if this statement is indeed correct, someone, somehow will be able to derive the proof. But how can we be so sure of this? Who can guarantee this? It is entirely possible that the statement is correct but that one cannot construct a proof (i.e., a finite sequence of logical arguments). This is the essence of Hilbert's challenge and the quest for mathematical

[9] Some set theorists might disagree and claim that the original intuition has been replaced by a "better intuition."

completeness. Every true statement should be provable. It might take "some time" to get the proof, but the proof should exist.

The vast majority of mathematicians believed this to be true and all they needed was a formal proof of this fact. Gödel took the challenge. At the beginning things looked just fine. Gödel was able to show that in some cases one can construct a system of axioms that is "bullet proof." Unfortunately, he also showed that these cases are relatively simple and without much structure or depth. More importantly, he demonstrated that as soon as one considers a system of axioms that introduces a sufficient level of complexity (something like the chess example), things quickly change - and not for the better. His famous Incompleteness Theorem states that for any self-consistent recursive axiomatic system powerful enough to describe the arithmetic of the natural numbers, *there are true propositions that cannot be proved from the axioms.* A few years later he built on his results and proved the Inconsistency Theorem. This theorem states that *one cannot prove that our arithmetic is consistent.*

Careful here, let me state this again: *He proved that one could never prove that arithmetic is consistent.*[10] Thus, the reality that mathematicians face is a bizarre one indeed: We could never prove that the arithmetic is consistent but we could prove that it is not! Namely, to disprove the consistency one only needs to produce a paradox. Thus, the whole world of mathematics faces a predicament similar to the one the set theory faced earlier on (remember, one cannot prove that something is a set, but one can prove that something is not a set).

Interestingly enough, another of Gödel's famous results deals with the Axiom of Choice and the consequences of its inclusion or exclusion from axiomatic theory. He could not prove that set theory (with or without the Axiom of Choice) is consistent but he managed to prove that the two theories (one with and another without this axiom) are either both consistent or both inconsistent. In other words, if one adds the Axiom of Choice, one does not make the things worse.

Before we continue with our quest, I have one more little twist to address. It is most likely just a coincidence, but one very uncanny coincidence, and I cannot help but include it here.

The year was 1901. As mentioned earlier, Newton's synthesis, which began in the mid-seventeenth century, created a strong bond between mathematics and physics, one that yielded a profoundly bountiful period for both fields. From that point, until the nineteenth century, both disciplines produced an astonishing array of successful theories, which not only revolutionized our world, but influenced each other as well. A large number of mathematical theories were inspired by physics and essentially all the physics theories were mathematical in their essence. However, by the mid-nineteenth century these two fields started to diverge, and this was mainly the doing of the mathematicians.

Nevertheless, although the initial level of cooperation reduced significantly, and although both disciplines acted more and more independently, they both

[10] "If a 'religion' is defined to be a system of ideas that contains unprovable statements, then Gödel taught us that mathematics is not only a religion, it is the only religion that can prove itself to be one": John D. Barrow, *Between Inner and Outer Space: Essays on Science, Art and Philosophy* (Oxford University Press, 2000), Part 4, ch. 13.

encountered a shocking challenge at the very beginning of twentieth century and in the very same year. As fate would have it, in 1901 Russell published his paradox and Planck his work on quanta. Both results profoundly changed their respective disciplines. Needless to say, both disciplines (mathematics and physics) were for the most part completely indifferent to each other's troubles.

But it did not stop there. After some time both disciplines adjusted to the new reality and both disciplines embraced the challenge. What followed was a steady flow of some of the most remarkable discoveries and astonishing developments that fundamentally changed both worlds: of mathematics and of physics. It took around 30 years before we got the verdicts. Remarkably, both decrees were delivered within just a few years of each other and both were disturbing and unusual. Not surprisingly, the mathematician and the physicist in question were only marginally (at best) aware of each other's work, but yet again both results sent a very similar (chilling) message and in very similar terms. Judge for yourself.

Heisenberg's Uncertainty Principle (1927) states that nobody can measure, arbitrarily precisely, the position and the velocity (momentum) of a particle simultaneously. These two quantities, position and velocity, are the focal points of Galileo's and Newton's work, and consequently this result touches the very core of classical physics, and in a very strange way. This inability to precisely measure things is not related to our intellect nor to our poor tools. It is just mathematics.

Gödel's Incompleteness Theorem (1930) states that one cannot prove that our axiomatic world is consistent. This is the fatal blow to Euclidian school in which everything starts with the axioms. Again, this inability to prove consistency is not related to our intellect nor to our talent. It is just mathematics.

6 On Postulates and Axioms

> If there is a God, he is a great mathematician.
>
> —Paul Dirac

We have learned a few things by now. To start with, we now appreciate how hard it is to create a consistent, self-sustained, and yet interesting universe. We have also learned the way scientists operate: They observe our universe (which seems to be self-sustained and interesting) and they try to reconstruct its laws, which they call *postulates*. Similarly, we have learned mathematicians' *modus operandi*: They create their own rules, called *axioms*, and then explore the universe that arises from these rules. In the following few chapters we will connect the two concepts - the axioms and the postulates - and try to unravel this mystery.

6.1 Theoretical Physics

Theoretical physics is a rather unusual endeavor. It falls somewhere between science and mathematics. While some would argue that theoretical physicists are both scientists as well as mathematicians, others would disagree and say that they are neither. Indeed, if we think of scientists in Galileo's terms, as people married to experiment, data, and fieldwork, people who roll up their sleeves and collect samples, be they rocks or insects, then theoretical physicists are not scientists. In this regard, they are much closer to mathematicians.

But, then again, theoretical physicists cannot be mathematicians for there is a profound difference at play. Yes, it is true that theoretical physicists, just like mathematicians, start with a few basic (mathematical) rules, and then, just like mathematicians, they derived their formulas and their theories. It is also true that they use mathematics, and essentially only mathematics, to do so. Nevertheless, there is a fundamental distinction between the two. Physicists care deeply whether their theories are in agreement with Mother Nature. Mathematicians do not.

Allow me to elaborate. Although only a small number of theoretical physicists would actually perform an experiment, essentially all of them *are acutely aware of experiments performed by others.* And they make sure that their theories are in agreement with these experiments. For any theory, no matter how beautiful and mathematically sound, if it is contradicted by an experiment it is automatically discarded by physicists. Mathematicians, on the other hand, do not discard their theories based on experiments.

Remember the story of Euclid's Fifth Axiom? Apparently, as we have learned, there were several geometries and they all seemed consistent. So, the logical (and urgent) question among the physicists was: *Which geometry is the correct one?* Makes sense, doesn't it? People wanted to know which of these geometries was applicable to our universe (thus the "correct" one). For mathematicians, this was not an issue. For us, they are all correct geometries. And to be fair, neither Euclid nor Lobachevsky ever claimed that their axioms were correct. All they claimed was: *If we assume that axioms are correct, then the following theory is correct as well.* Nothing more and nothing less.

So, in many ways, compared to mathematicians, theoretical physicists have a much tougher job. Not only must their postulates be right, not only must they correctly develop and interpret any theory that arises from these postulates, but they must also confirm their findings with an experiment. Nothing remotely similar is required of mathematicians. Our theorems are not scrutinized by experiments and our axioms are not questioned for their correctness. Thus, we must salute theoretical physicists, for their job is truly an extraordinary one.

And there is no better story to prove this point than Newton's. His persona is well known and renowned, but I cannot help but observe that much of the "folklore" associated with him is somewhat misguided. And I wonder how many people truly understand what it was that he actually accomplished. For you see, what he did is nothing short of magical. Allow me to elaborate. Newton's three laws (postulates) are:

Postulate 1. An object either remains at rest or continues to move at a constant velocity, unless acted upon by a force.

Postulate 2. The vector sum of the forces F on an object is equal to the mass m of that object multiplied by the acceleration a of the object: $F = ma$.

Postulate 3. When one body exerts a force on a second body, the second body simultaneously exerts a force equal in magnitude and opposite in direction on the first body.

Let us pause here for a second. Let us try to understand what the postulates say. The third one is the "weirdest." It is often referred to as a postulate "on action-and-reaction," and we have all heard it, many a time, ever since grade school. But do we really understand it? Our teachers would say: *When you step on the floor, you feel the force on your foot, and so does the earth; it feels the same force acting on it.* What a "silly" explanation. And, even if correct, why would this particular mechanism, seemingly so insignificant and strange as well, be declared as one of the three most important laws of physics?

That was the third postulate. But the first one is even harder to accept. To start with, it is wrong. Or at least, it looks wrong: *Any object, if left alone, will eventually stop*. Everybody knows that. Just kick a ball and wait until it stops. And yet, here is a postulate, a fundamental law, which states that the opposite is true. But what about a vacuum or a frictionless universe, one might ask? So, what about them? Newton (and Galileo before him) had no idea about the existence of a vacuum and could not have performed the experiments we see nowadays. So how on earth did he (they actually) come up with this postulate?

And then comes the easy one, the second postulate: Force applied to an object will cause the object to move (accelerate). Finally, a postulate that is easy to comprehend. We have all experienced it: The larger the object (larger mass m), the larger the force necessary to move it. And all this is captured with an elegant equation: $F = am$. Perfect.

So there you have it, Newtown's three postulates. But where is the magic which I promised a few lines earlier? Well, let us see: It was the seventeenth century and the times were exciting. Sailboats were bringing goods from the far ends of the earth, horse carriages were crisscrossing the countryside. People were moving, as were ducks in a pond and whales in the ocean. And the oceans, too, were moving, bringing catastrophic storms at random times. Cannonballs were flying and birds too. All so chaotic, so different, and so beautiful. And here comes Newton who essentially declares: "All these things you see around, the ducks, the whales, the oceans and the cannonballs, they all obey my three laws. And yes, two out of these three laws do not even make much sense, and yet, here I declare, that all that moves and will ever move, must follow these three laws. And no other law."

What a bombastic idea! And yet, he was right. Which, to be honest, is nothing short of magical. For you see, this was the first time in recorded history that someone had discovered (or guessed?) some of the "creator's" axioms.[1] And although it is astonishing that his three postulates are correct, far more surprising is the fact that there are only three of them. Who would have thought that everything that moves around us is fully described by just three rules? Our intuition, our observations of this complex and chaotic world around us, would have suggested dozens (if not hundreds) of rules necessary to describe it. But no. There are only three.

6.1.1 A Few Simple Rules

The idea of *a few simple rules* did not stop with Newton. We have already seen that Einstein used only two postulates for his special theory of relativity. Thus, everything we have ever learned about the strange world of relativistic mechanics, time that

[1] "If I have seen further it is by standing on the shoulders of Giants," is what Newton said (see "Letter from Sir Isaac Newton to Robert Hooke," Historical Society of Pennsylvania, 1675, digitallibrary.hsp.org/index.php/Detail/Object/Show/object_id/9285.Google Scholar_. Indeed, it appears that Galileo had already devised two of Newton's three laws. Descartes, too, had his own three laws (albeit the third one was wrong), and it appears that Huygens (a few years before Newton) had come tantalizingly close to discovering the laws himself.

slows down, the twins paradox, the mass that increases, all of it, is a mathematical consequence of just two simple rules. Planck's example concurs. He actually added only one postulate, on the minimum amount of energy, which created a chain reaction resulting in its own avalanche of bizarre results.

Throughout the last few centuries, ever since Newton, we have all been indoctrinated by this idea: *a few simple rules.* Generations of us have been exposed to science, and although we all know that learning science is a rather complex endeavor, we also know that deep down, at their essence, the fundamental rules of science are simple and not very numerous. The law of gravity is captured by an almost unbelievably simple formula. So is magnetic force, and so is kinetic energy, and so is centripetal force, and so on. All life on earth, fungi and bacteria, whales and giant sequoias, they all follow from 20 peptides, coded by just four amino acids. Atoms are explained with a model that high-school students can understand. The theory of evolution is so simple that grade-school children can comprehend it.

We are so comfortable with this notion that we often forget how very recent it is. Through the eons, literally, from caveman until Newton, humans were under the completely opposite impression. *A few and simple rules? Absolutely not!* For centuries before Newton there were countless clever men, and they all have observed our world and they all concluded the same thing: *Yes, our world seems to obey certain rules but there are so many of these rules and they all seem so complex and hard to understand.* These wise men, whether in ancient Egypt or China or India, had a doctrine that was the complete opposite of the idea of a *few simple rules.*

Just recall the convoluted Greek constructions with glass spheres which were supposed to explain the movement of the stars. Or the intricate Chinese zodiacs, or the earth on the back of a giant turtle, or the infinitely complex Greek and Romans pantheons with myriad of gods, each responsible for his or her own domain of natural world.

But, novel or not, this notion of a few and simple rules seems to be correct. By and large, scientists are now convinced that our universe is indeed based on some incredibly simple and not very numerous postulates. Some of these postulates were discovered by a pure stroke of genius. Newton's and Einstein's examples fit this category. Some of these postulates were discovered theoretically, with no small help from mathematics. Planck's and Heisenberg's examples fit this category. And many other postulates were discovered through combination of experiment, theory, common sense, and lot of hard work: Darwin's Evolution Principle, Watson and Crick's DNA model, Mendeleev's periodic table – the list goes on.

But why just a few simple rules? Why not thousands of rules? Why not very complicated and very hard-to-comprehend rules? Newton's second law is captured by the trivial equation $F = am$, but why not $F = (a^3 + a)e^{m^2}$? This second equation, although considerably more complex, also captures Newton's intuition – *large mass acceleration requires large force.* So why was nature so kind to us and decided to bestow us with the first (trivial) formula instead? Einstein himself was puzzled, for he said: "The most incomprehensible thing about the world is that it is comprehensible."[2]

[2] From "Physics and Reality" (1936), in *Ideas and Opinions*, trans. Sonja Bargmann (New York: Bonanza, 1954), p. 292.

Since our universe is comprehensible mainly because it is based on just a few and simple rules, Einstein is essentially asking the same question as we are: Why only a few simple rules? And he is the right person to ask, since he used only two postulates to describe his special relativity theory. Had the universe been "malicious" and required not two but 202 postulates, Einstein would have never been able to correctly decipher the relativity theory. It would have been incomprehensible to him.

So, why is the world comprehensible? Well, believe it or not, we have already answered this question. All the pieces of the puzzle are in front of us, and only some assembly is required. But, before we start "assembling," a few more lines are in order.

6.2 Mathematicians and Their Axioms

We start with an anecdote (this time a true story). This was a few years back, in a workshop on Probability on Banach Spaces (probability on infinite dimensional spaces). There were some 50 of us at a remote institute, with very limited internet connection, to talk mathematics and (hopefully) socialize. The former worked well. After dinners, during afternoons and evenings, some of us would come down to the big living room, but we would mainly sit alone and would converse only occasionally and awkwardly; mathematicians being largely introverts. But not Vlad and Goran. The two of them could be seen sitting together at the corner table for hours and hours, deep in discussion and scribbling something on paper. It was a curious sight, so after a few days I had to ask:

> Hey guys, what are you working on?

A big mistake. All too eager they started explaining their stuff, and all too quickly I was completely lost. Although we belonged to the same sub-sub-sub-subfield of mathematics, it took me only 45 seconds to utterly lose them. All I remember (to this very day) are the words "Kolmogorov" and "stochastic differential equation." But my confused face, and my facial clues, and my "funny" comments, all designed to communicate that I did not follow, were lost on them. Oblivious to any social norms, they went on explaining their problem in ever-so-minute details. It took me a good 20 minutes, a very painful 20 minutes, to extract myself from this awkward (and one-sided) conversation.

Fast-forward three years (we meet every three years), this time in New Mexico, at Santa Fe Institute. Same group, same format, and what do you know? Vlad and Goran sitting together in the corner, scribbling for hours, every day. Since I had learned my lesson I avoided them in large circles. But, then again, since I like a good joke, I could not resist, and one day, from a safe distance, I had to tease them:

> Hey guys, are you still working on the same equation?

The look they gave me. Instantly I knew I had said something wrong for they stared at me as if I were Bozo the Clown. Then Goran spoke:

> Of course we are working on the same equation. What? You actually thought we would have solved it in just three years?

-----------------<>-----------------

There you have it. Vlad and Goran are by no means unusual. For better or worse, much of mathematics is done this way. Lobachevsky had spent innumerable hours developing his geometry, which went against intuition and frankly any common sense. Abel devoted a good portion of his adult life proving that one *cannot solve* equations of the 5th degree (and higher), an astonishing result but without any impact on science whatsoever. Weierstrass, too, had meticulously worked on his thorny and very complicated function, one that is continuous but nowhere differentiable. And for what purpose?

Andrew Wiles basically locked himself away for seven years and finally proved Fermat's Last Theorem (after 360 years of futile attempts by so many before him). Again, an astonishing result but completely insulated from any trace of physical reality or applications. Grigori Perelman proved the Poincaré conjecture (after 100-plus years of efforts and many dead ends that preceded him) but refused the Fields Medal, stating: "I'm not interested in money or fame; I don't want to be on display like an animal in a zoo."[3] Any mathematician skilled in the art could easily recite dozens and dozens of similar examples about well-known mathematicians, and probably hundreds of personal accounts as well.

Thus, for better or worse, we mathematicians (*for definition please see the disclaimer at the beginning of the book*), are an unusual bunch. Unlike scientists, engineers, and others, we typically work alone. We do not need labs or fancy equipment; we do not require expensive mass spectrometers or solid state lasers. Our art, not unlike that of poets, requires only a pencil and some paper. And yet, our art, it seems, runs the universe. Who would have thought?

We have already quoted Einstein: "How can it be that mathematics, being after all a product of human thought independent of experience, is so admirably adapted to the objects of reality?"[4]

Well, not so fast. At first glance, Einstein was right. Much of mathematics is done by people who rely solely on their thoughts, on their human minds, and who work independently of experience. This is all true, since most mathematicians do not "inconvenience" themselves with physical reality, experiments, and such. Nevertheless, there is one crucial and fundamental flaw regarding this premise. It has to do with the axioms.

Even if one accepts that all mathematicians are like Wiles and Perliman, locked in their attics, isolated from the world, working on their problems, and even if one assumes that all of mathematics is done this way, one must admit that the axioms used by these mathematicians are very much in tune with our physical reality. Thus, while the theories we develop might be a *product of the human mind*, the axioms from which mathematics was created are very much products of the world around us. Of that, there is very little doubt.

We start at the beginning: Euclid. If one observes his five axioms, one cannot but agree that they are inseparable from our experience and very much influenced by an

[3] Russian math genius Perelman was urged to take $1m prize: BBC news. March 24, 2010, https://en.wikipedia.org/wiki/Grigori_Perelman.

[4] From A. Einstein's lecture, "Geometry and Experience," given on January 27, 1921 at the Prussian Academy of Sciences in Berlin. The English translation was published by Methuen & Co., London, in 1922.

experiment. While it is not easy to find a straight line or a perfect circle in nature, these two, the line and the circle, are essential to any engineer or architect. If one takes a closer look at Euclid's Axioms, one easily recognizes an experiment. Namely, the axioms capture the exact procedure needed to draw the circle and the line (using a ruler and compass). A mathematical purist would claim that the "line" and the "circle" in these axioms need not be the actual lines and circles but any objects satisfying the axioms. While this is a correct claim, it does not negate the fact that these axioms are obviously inspired by our experiences and the physical world around us.

Euclid's Axioms are not special in this regard. In the Post Scriptum, at the end of this book, we will show that a large portion of mathematical axioms did not appear out of "thin air," but are very much motivated by physics and technology. Here, we will argue the same point, but using a much simpler and easier to follow case.

6.2.1 Real Numbers and Their Axioms

To start with, we must notice that something "funny" is going on here. The very name *real numbers* is strange one. Are there any unreal numbers? Also, we have *rational* numbers (fractions), *irrational* numbers (nonfractions), and *imaginary* numbers. After many years of schooling we became accustomed to these labels, but when an intelligent person encounters them for the first time they all seem so very strange. As a philosophy major once asked me: *Why would a number be reasonable (rational) or unreasonable (irrational), real or unreal?* Well here is why.

Mathematicians axiomatized the idea of real numbers using 13 axioms, 12 of which are reasonable.[5] Namely, the first 12 axioms are as clear as day and very much inspired by physical reality, specifically geometry. For example, the commutative property $ab = ba$ is obviously related to the area of a rectangle. The axiom captures the simple experience: If we flip a rectangle on its side we do not change its area.

The two rectangles have same area, thus the axiom $ab = ba$ (see Figure 6.1).

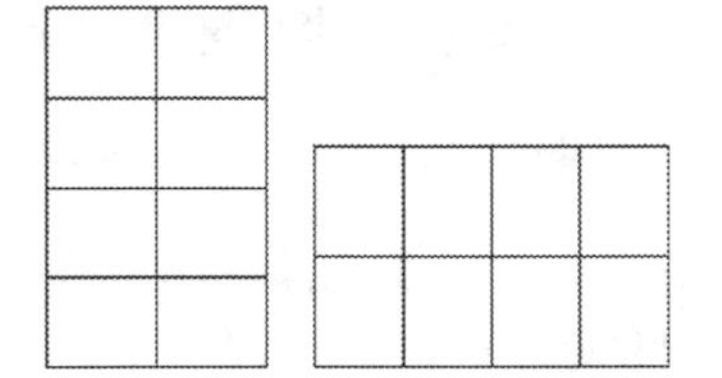

Figure 6.1 Commutative property.

The distributive property $a(b+c) = ab+ac$ is particularly interesting. The reader has been exposed to this rule many a time, and she probably does not realize how tricky it is. To start with, it is this (and essentially only this) axiom that students have a hard time with (factoring.). But also, if one were to justify this rule to a student, one realizes that it is not that easy. One cannot prove it (it is an axiom), and the only way to argue it, is to use geometry (see Figure 6.2). This axiom is intrinsically connected to some very specific, earthly objects and not to abstract reasoning.

[5] Some books group the axioms differently, so the 13 axioms sometimes become 15 or 16 axioms.

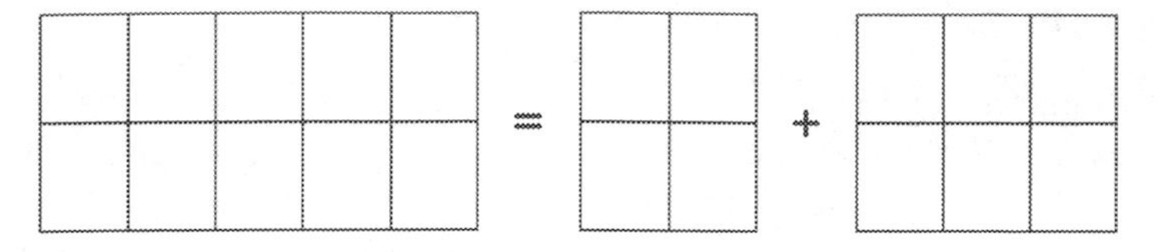
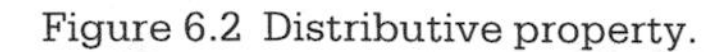

Figure 6.2 Distributive property.
Note: Justification for the axiom: $a(b+c) = ab+ac$ (the areas are equal).

Similarly, we could argue that each of the first 12 axioms is inspired by physical reality and our experiences. But not the 13th axiom. That one is a product of our thoughts, and very much not inspired by physics. In many ways it runs contrary to physical reality, as well as to experiments.

More precisely, the 13th axiom states that any bounded set must include the smallest bound, also called *supremum*. An example will help.
Take a set:

$$\boldsymbol{A} = \{1, 1.9, 1.99, 1.999, 1.9999,. 1.99999,...\}.$$

In other words, our set consists of all decimal numbers that start with 1, followed by a decimal point, followed by k digits of "9," where $k = 0, 1, 2, 3, 4$, etc. Clearly, our set $\boldsymbol{A}$ is bounded, since every element in $\boldsymbol{A}$ is strictly smaller than the number 2. Also, it is clear that 2 is the smallest of such bounds. Thus 2 is a *supremum*, or a smallest upper bound. Easy, isn't it?

Well, not so fast. The axiom states that *every bounded set* must have the *supremum* (it is unique). However, the axiom does not specify which real number is the supremum of a given set, nor does it describe a procedure by which to compute such a supremum. No, it only commands that such a supremum must exist. Consequences? Motivation?

I am glad you asked. It turns out that without this axiom we could not define convergence. And without convergence we do not have limits, and without limits we do not have derivatives and integrals, and without them, we do not have calculus. And without calculus, well, frankly, Einstein would not have been surprised about power of mathematics, since it wouldn't have been there as a large portion of mathematical intrusions into the world of physics are, in one way or another, related to calculus.

So, here we have it. The conclusion of this chapter: Yes, many, if not most, mathematical axioms are inspired and influenced by our experience and physical reality. But not all of them. There is a substantial part of mathematics that indeed comes *from our thoughts, independent from experience.* In fact, the 13th axiom contradicts quantum physics.[6] It goes against our experience, against experiment. And yet it is responsible for a wide range of mathematical intrusions into the world of physics.

[6] This axiom gives a rise to irrational numbers which are not always compatible with quantum restrictions.

6.3 The Rare Axioms

The time has come to unravel this mystery, about the intrusion of mathematics into the world of science. In what follows I will present a plausible solution. Unfortunately, that is all I can do. We are not talking about mathematics here so I cannot offer a proof, nor are we dealing with physics so I cannot offer an experiment to justify my "solution." We are dealing with philosophy, so the best I can do is to offer a logically sound argument, a "plausible" solution. Here it comes.

6.3.1 An Alien Board Game - Again

In this analogy, we spoke of a fictional spaceship, game players, and followers of a "certain" cult, and we connected these to the way scientists and mathematicians operate. At the end, we concluded the following:

> The rules of our universe are based on a small set of relatively simple, noncontradictory axioms.
> Mathematicians have been studying all kind of theories based on a small set of noncontradictory axioms.

Now, it does not seem particularly surprising that some mathematical theories are in perfect harmony with the rules scientists encountered studying our universe. In particular, since mathematicians had 2,000-year head start.

That was a conclusion, and here is the critique:

a. Why would our universe be based on a few simple mathematical rules? Why not on a few thousand very complicated mathematical rules?
b. There are infinitely many fundamentally different axiomatic theories, and 2,000-year head start does not seem like much. It is still extremely unlikely that theories designed by mathematicians are in agreement with our universe.

In preceding chapters, we managed to answer some of the critique.

Counter - a. We have shown that not only is our universe based on a few simple rules, but that it had to be that way. The same is true for any mathematical theory - it had to be based on a few and simple rules. Why? Because any game, any axiomatic system, any universe which does not follow this principle collapses very quickly.[7] In other words, numerous and complicated initial rules yield a contradictory game.

Counter - b. Yes, technically mathematicians could have studied any of the infinitely many axiomatic systems, regardless if they are based on physical reality or not. But they did not. We have shown that the vast majority of mathematical axioms are very much inspired by the physical world around us. This fact, coupled with the 2,000-year head start, now significantly increases the chances that some mathematical theories are in perfect harmony with the rules scientists encountered studying our universe.

[7] Or at least the vast majority of them collapse very quickly.

But there is more. We have also learned that vast majority of universes (games) which are based on a few and simple rules are very boring. In other words, it is rare we get a perfect match: a few simple axioms that also manage to create a complex and noncontradictory universe. Our universe is one such a match and so is our mathematics.

The Rare Axioms Hypothesis

Mathematicians study any simple axiomatic system that yields a rich and consistent theory; however, such theories (universes) are very rare. Our universe is one such system, and mathematical axioms are influenced by our universe. The intrusion of mathematics into the world of physics now follows easily. All that is needed is for mathematicians to derive logically correct theories from the right axioms. And that is what we do.

PART II

The Oracle

7 Introducing the Oracle

7.1 A Time to Recall and a Time to Resume

Our "Rare Axioms Hypothesis" seems spot on. First, it makes sense. The whole premise does, and one gets that feeling, well known to any scientist or mathematician, when so many pieces of a puzzle begin to interlock. *This fits so well, it cannot be just a coincidence.* And if one investigates the main premise in some detail, it gets even better.

For example, this revelation calls for the *Shotgun Principle.* In other words, since mathematicians had a 2,000-year head start on scientists, and since we have been investigating all kinds of axiomatic structures, regardless of their application, it is logical to expect that the vast majority of our results would have nothing to do with nature and the laws of our universe. In other words, mathematicians are firing a shotgun with thousands of little pellets (theorems), of which only a select few hit the target (applicable results). That is, one expects that only a small portion of mathematical theorems would embed themselves into the fabric of our actual universe and its laws. And, as it turns out, this is an easily demonstrable truth.

For example, in Chapter 1 we talked about "amicable numbers," the ones for which the sum of all divisors of one number is equal to the other (e.g., 220 and 284). Amicable numbers have preoccupied mathematicians for millennia and yet there are no applications, direct or indirect, related to this result.[1] In fact, all of number theory – one of the most ancient and distinguished branches of mathematics, second only to geometry – is riddled with useless (but beautiful) theorems. We have already mentioned the Goldbach conjecture and the claim that every even number

[1] John Conway allegedly said that the only application or use for these numbers is the original one – you insert a pair of amicable numbers into a pair of amulets, of which you wear one yourself and give the other to your beloved.

can be expressed as the sum of two primes. This statement, too, attracted scores of mathematical heavyweights, and yet the conjecture has remained unsolved for few centuries now. Needless to say, it does not seem to be very useful.

Fermat's Last Theorem is another example. The conjecture was that it is impossible to find integers such that $x^n + y^n = z^n$ (for $n>2$). This was proven, some 360 years after it was postulated. The proof was incredibly complex. It fused very distant and intricate branches of mathematics (elliptic curves and such), and, needless to say, provided zero insight into the laws of nature, or any other application to science and engineering. And yet, "Fermat's Little Theorem," the smaller sibling to Fermat's Theorem, which also talks about integers and primes, is an instrumental tool in modern cryptology, which thrives on number theory.

This little sojourn into number theory and its role in the modern world fits perfectly with our revelation, as it reinforces the *Shotgun Principle*. That is, mathematicians provided a massive theory, with tens of thousands of theorems, lemmas, and propositions – the vast majority of which are completely disconnected from science and technology – but with a few pellets (theorems) that hit the target perfectly. Any person skilled in the art could produce many similar shotgun examples, among different branches and sub branches of mathematics. My research is related to a particular sub-subfield of mathematics. It deals with probability on infinitely dimensional spaces, a topic not much related to our three-dimensional world. Consequently, the vast majority of results published in this area have nothing to do with science or technology. And yet some of these results embedded themselves in the very fabric of modern statistics and computer learning applications. Some of the pellets hit the target. Thus, it seems that the "Rare Axioms Hypothesis" is confirmed.

And yet, I am skeptical. Here's why.

7.2 A Critique of the Rare Axioms Hypothesis

7.2.1 What are the chances?

The very first mathematical intrusion into the world of physics presented here, and in my opinion the most striking one, is about the ellipse and the earth's trajectory. The Shotgun Principle would suggest the following scenario: After Euclid's Axioms, mathematicians would have developed their theories and, in the process, they would have described a myriad of different curves and trajectories. They would have continued like this for a thousand years without any idea that some of these curves might be used by Newton. Much later, when the law of gravity was discovered, one would expect that among untold number of previously described curves (shotgun pellets), one curve (one pellet) would be the earth's trajectory. At the first glance, that is exactly what had happened.

But there was a curious little twist! True, the ellipse was one among the myriad curves that had been studied before Newton's time, but it is also true that the ellipse was *the very first* nontrivial curve ever studied. For you see, once the ancient Greeks began their geometric adventure, they started with the objects stipulated among Euclid's Axioms: the line and the circle. For a few hundred years they studied these simple curves, and after they proved essentially all that was worth proving in this setting they moved on. They entered the kingdom of more complex curves. But then this wacky thing happened: The very first curve they considered was the one that happened to describe the earth's trajectory around the sun.

This is completely contrary to the Shotgun Principle, for it seems that the very first pellet that was ever fired by the geometric shotgun found its target. What are the chances of that? How is this possible?

Fortunately, mathematics is not only one of the oldest human activities it is also one of the best-documented human activities. We have a very good idea of how, why, and who, studied these curves, so maybe a closer look will shed some light on this magical hand that guided the early geometers toward the prophetic ellipse. Let us see.

Everything boils down to a curious little logical game practiced by the ancient Greeks. They did not play chess or Go, but they liked the following puzzle: Using only a compass and a ruler construct a particular geometric structure; the more complex the construction the better the puzzle. With these two instruments one can easily draw a circle and a triangle, but one can also construct a right angle, perform angle bisection, find the center of mass, and much more. In fact, there are hundreds and hundreds of these puzzling constructions. Needless to say, the whole exercise has very little to do with reality and the way our universe works. Nature does not limit itself to a compass and a ruler.

Nevertheless, this was a very popular pastime for many Greek intellectuals, not unlike chess or bridge or the word puzzles one finds in Sunday newspapers. These were very interesting puzzles that involved some very specialized skills. Over time, as more and more constructions were discovered, some hard problems started to emerge. For example, using only a compass and a ruler it is easy to construct a hexagon, but it turns out that it is rather hard to construct a pentagon (the reader is welcome to try). The most famous of these hard problems were: squaring a circle, trisecting an angle, and doubling a cube. The last one deserves our attention.

The problem is as follows: Given a cube with side a and volume V, construct a cube with volume $2V$ (in other words, double it). This is a trivial problem since all one needs to do is to compute $\sqrt[3]{2}$ (obviously, $V = a^3$ and the desired cube should have the side $a\sqrt[3]{2}$). But the rules of the game forbade any computation. One must perform this doubling only by means of a compass and a ruler. Why the restriction? Well because those were the rules of the game. Why is the knight able to jump over other figures in game of chess? Because those are the rules.

I hope the reader realizes how distant, how unrelated to any axioms of our universe the whole exercise was. There is nothing here that even remotely resembles Newton's laws or any other laws of nature.

To add insult to injury, mathematicians were able to prove (about 2,000 years later) that this problem is not only unrelated to any physical reality, it is unrelated to any mathematics as well: It is unsolvable. The solution does not exist. It was a dead-end problem.

So how does this connect to our story? Well, it turns out that one of the many who attempted to solve this problem was the mathematician Menaechmus. In his futile efforts to double the cube he considered the cross sections of a cone and a plane. Since the problem is unsolvable the whole attempt failed, but in the process he discovered these three curves: the parabola, the hyperbola, and the ellipse.

Think about it. These curves were byproducts of a dead-end attempt at an unsolvable problem generated by an ancient game. Clearly, these curves were just random pellets fired from the geometric shotgun. And yet they are the very first complex curves that humans studied and they are intrinsically related to one of the most fundamental laws of nature: the law of gravity. So, dear reader, I hope you agree that this story very clearly demonstrates that the Shotgun Principle did not play a role here. What are the chances that the very first random pellet will strike the target? And there are so many similar examples.

If the reader will recall, the next example we introduced in this book was about angles and music. There, we explained the rise of trig functions; this mathematical intrusion also contradicts the Shotgun Principle in some very fundamental ways, as we shall see shortly.

Recall that the science of waves and thermodynamics are some of the pillars of classical physics. These two phenomena have very little in common; one is mainly experienced through sound and vibrations, while the other through heat and temperature. Consequently, these two fields were long considered to be branches of science distinct from one another. Nevertheless, once physicists learned how to properly mathematically treat these two phenomena it became clear that they have quite a bit in common. But the connection was mainly mathematical and not physical. Both fields may be described in terms of trigonometric functions.

Quantum mechanics, a field that has very little philosophical or logical connections to any of the human intellectual activity that preceded it, suggests many bizarre laws of nature. One of the most striking is the wave–particle duality, which states that all objects are simultaneously material objects as well as electromagnetic waves. And a wave, as we know, is mathematically represented by trigonometric function. Thus, the simple trick of Sin(x), the ratio of triangle's side and hypotenuses, somehow embedded itself in so many fundamental scientific concepts. By the sheer volume of theory that followed, this trigonometric intrusion dwarfs the one of the ellipse.

But if one applies the Shotgun Principle to that of the waves–thermodynamic-quantum mechanics saga, one would expect the following scenario: Physicists, after discovering the importance of waves, would turn to mathematics and hope that among the thousands of curves and functions in there they would find a curve that would be able to describe this phenomenon: the wave. And, naturally, they would hope that the function in question would be simple and expressible in some elegant form. Incredibly, physicists got lucky, since the curve in question exists, and is not represented by some monster formula but a very elegant function: $f(x) = \text{Sin}(x)$.

Well, it turns out that the functions Sin(x) and Cos(x) belong to a class of nonalgebraic functions (nonalgebraic functions are not expressible using the standard algebraic operations of multiplication and addition). As one would expect, through the centuries mathematicians had described hundreds of these nonalgebraic functions. But, in another violation of the Shotgun Principle, the very first nonalgebraic function humans ever studied (the trig function) happened to involve one of the most important (and the most applicable) of them all. Thus again, the very first pellet fired from this nonalgebraic shotgun hit the target perfectly. How is this possible?

We have a peculiar situation here. It seems as if some invisible force has guided mathematicians to do the "right thing," as if some kind of mathematical oracle has pulled the strings. It seems that this oracle could take an engineer working on thermodynamics, an acoustic scientist working on modulations, and a theoretical physicist working on quantum mechanics, and somehow unite all three of them with ancient geometers and their triangles as if some vortex pulls unrelated concepts from vastly different backgrounds toward some common, often ancient, mathematical theory.

Not all math concepts are granted this status but a surprising number are. Our fourth example, the tale of *flowers and spiky curves*, is one in which mathematicians deliberately created a monster, a curve that was as far from real-life applications as one could possibly be (as Poincaré so eloquently put it). But even this case was not immune to the charms of the mathematical oracle. Indeed, this nowhere differentiable continuous curve (i.e., spiky curve) is realized by the Brownian motion, which was the first continuous stochastic process ever studied. And here we go again: The very first pellet fired by the stochastic gun is fundamentally connected to atoms and their paths, as well as to the behavior of stock market. What are the odds? (Pun intended.)

7.2.2 Why simple formulas?

These are indirect arguments, but we can do better, we can argue directly. Namely, even if we ignore the aforementioned critique of the Shotgun Principle, we still have an unfathomable problem to address.

We start with a simple observation: Mathematics, once applied to the world of physics, very quickly, astonishingly quickly, becomes cumbersome, complicated, and all but intractable. I am sure the reader has encountered cumbersome mathematics before, but it pays to revisit it.

Let us start with the simplest curve we can think of, the parabola, and let us try to compute its length. This is not only a reasonable mathematical question but also an engineering one. And while the curve is elegant and the question is easy, the answer is the following monster:

Length of parabola formula: $y = x^2$

$$f(x) = \frac{1}{4}\ln\left(2x + \sqrt{1+4x^2}\right) + \frac{1}{2}x\sqrt{1+4x^2}.$$

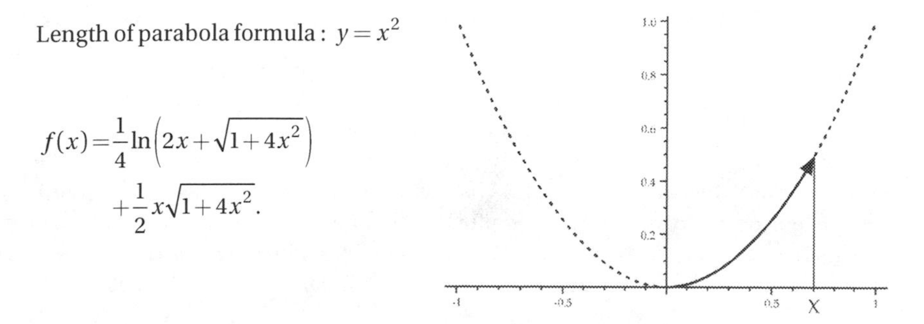

Apply the same question to another elegant curve. Take a wave and try to compute its length. That is, imagine a line that traces the top of a beach swell or the length that a sailboat must traverse as it sails on top of the swells. How long is this line tracing Sin(x) for a given x on the x-axis? This is an obvious question to ask, isn't it? But not an easy one to answer. In fact, it is impossible to write this formula (at least not by using finitely many terms).

$f(x) = ?$ Nobody knows.

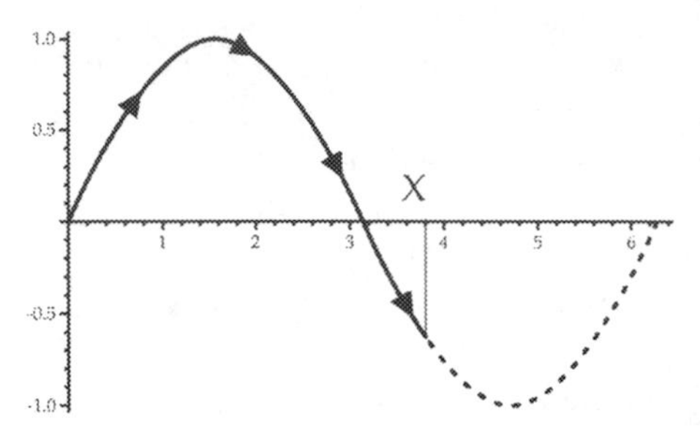

Or take the celebrated "bell curve," the one that appears in so many statistical applications. This curve resembles the path traced by a simple formula: $f(x) = 1/(1+x^2)$. Indeed:

$$f(x) = \frac{1}{1+x^2}.$$

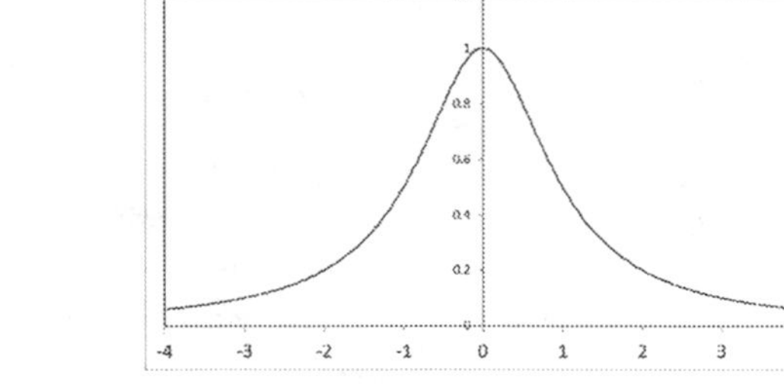

However, nature chose the following ogre instead:

$$f(x) = \sqrt{\frac{1}{2\pi\sigma^2}}\, e^{\frac{-(x-\mu)^2}{2\sigma^2}}.$$

Or, let us recall oscillations and trig functions. Yes, it is true that they are connected, but not always very elegantly. Imagine a child gently swinging back and forth on a swing at a playground. The formula that captures this motion is this "monster":

$$f(x) = e^{ax}\,\mathrm{Sin}(x\omega + \varphi) + e^{bx}\,\mathrm{Cos}(x\omega + \gamma).$$

And there are many, many more. Thousands more. I am sure the reader is not surprised, and she agrees: Applications of mathematics are very often very complicated. In fact, as soon as we move even slightly toward mathematical descriptions of the physical world, we encounter mathematics that very quickly sees elegant axioms transform into intractable and hard-to-follow concepts.

I hope the reader sees the point I am trying to make, and why this notion is in conflict with the Rare Axioms Hypothesis. Namely, even if we start with simple mathematical concepts, once we apply them to the world of physics we very quickly encounter very complicated formulas. Thus, Rare Axioms or not, it is unreasonable to expect simple, ancient mathematical concepts to play any role in modern science and technology. They should have entangled themselves into complicated and intractable formulas centuries ago. But they did not.

We finish this chapter with an example, one that is at the same time utterly confusing and encouraging. It is an example that offers a glimpse of hope, for it points toward the solution of our puzzle, if only we look carefully.

7.2.3 *The hanging rope problem.*

Take the following geometric/engineering construction: a rope hanging between two buildings or pillars. This particular technique, of stretching a string or a rope during construction, is as old as engineering itself (and is very much in use today). With near mathematical certainty we can claim that the ancient Greeks had encountered this situation. But a rope hanging between two buildings will not stay straight, no matter how hard we pull. It will curve. It will sag in the middle, pulled by its weight and gravity.

Thus, again with near mathematical certainty, we can claim that this particular curve was observed by the ancient Greek engineers as well as mathematicians. And here comes the issue: Since this curve presents the most obvious deviation from a straight line (a shortest line between two points), it is very likely that ancient Greeks had encountered it before any other curve. Likely even before circle itself. But they did not study it. And here is why:

$$f(x) = \frac{e^{-x} + e^{x}}{2}.$$

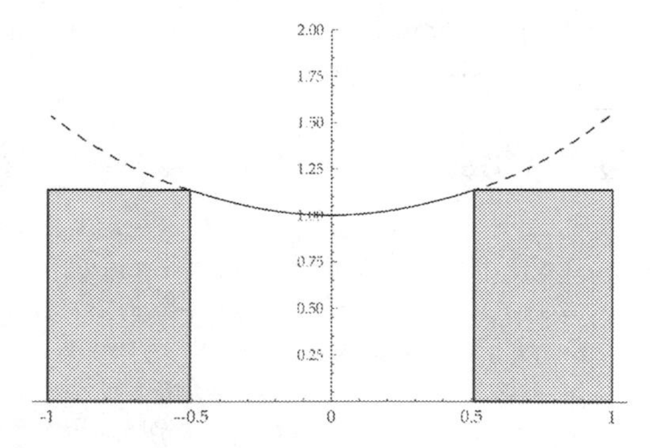

The formula is too complicated; it involves exponential (hyperbolic) functions, and the Greeks did not know how to deal with it.

So, what did the Greek mathematicians do? Well, just like the restless student who ignores a hard homework problem and doodles something fun and interesting instead, the Greeks did the same. The actual engineering problem was too hard, so they "doodled" something fun and completely unrelated. The engineers asked: How long should a rope between two buildings be?[2] but the mathematicians ignored that question and worked on intersections of a cone and a plane. Why? Because it was more fun.

[2] Another strange coincidence: The formula for the length of any curve is notoriously complicated, so much so that even for very easy functions we get impossible equations. However, the one function for which everything fits perfectly is the one that describes the length of the hanging rope. What are the chances?

The rest of this story we know all too well. The Greeks took the "easy way out," and instead of cumbersome hyperbolic curves they opted for ellipses, parabolas, and such. These curves were much easier to study and they connected nicely to the circles, lines, and triangles the Greeks knew so well. And this approach, this synergy, created a bonanza of new theorems, properties, and curves to study.

All true and all well documented. But how does it help with our quest? Well, for one thing, we just saw the oracle in action. We have just described a mechanism that nudged Greek mathematicians away from hyperbolic functions and toward conic curves. Let us describe this mechanism in some detail.

What exactly did the Greek mathematicians do? They took the path of least resistance. They studied the easiest curves they could find and the curves that mathematically interacted with other mathematics they knew. This particular scenario has been repeated countless times by countless mathematicians. In layman's terms: We all go with the flow. We all go where mathematics tells us to go. We all pursue theories that are easier to pursue and that connect to other theories we know. And that is the clue for our quest.

For you see, we mathematicians got lucky, incredibly lucky. This "flow" never seems to end. In mathematics there are myriad, maybe even an infinite, number of similar flows and paths, all somehow connected to each other. And all we have done, for the last 3,000 years or so, is to follow the stream. This very fact, about incredibly interconnected mathematics, is, dear reader, the very essence of my argument. Which leads us to the next section.

7.3 Down the Rabbit Hole

That mathematics is interconnected, that we keep seeing similar concepts appearing and reappearing throughout different textbooks and mathematical theories, well that is "old news." Everybody knows it. Ever since high school we have learned to expect π s and *e*s, as well as *Sin*s and *Cos*s popping up at random places and times. And the more one learns, the higher one climbs on the mathematical Mount Olympus, the worse it gets, for these bizarre connections only multiply and they only get stronger, and more bizarre and convoluted.

Nobody knows for sure when and how this all started, nor who was the first one to notice it. Nevertheless, for the sake of storytelling, we can argue that these two played a defining role:

$$\frac{\pi}{4}=\frac{1}{1}-\frac{1}{3}+\frac{1}{5}-\frac{1}{7}+.-\cdots, \qquad \frac{\pi^2}{6}=\frac{1}{1^2}+\frac{1}{2^2}+\frac{1}{3^2}+\frac{1}{4^2}+\cdots.$$

There is nothing in the definition of an odd number that could even remotely point toward the above connection to π. The same is true for the reciprocals of the square integers. And yet, the formulas are there, as real as the Rock of Gibraltar. And probably even more so.

Through the last few centuries, mathematicians have produced thousands (probably tens of thousands) of such formulas and concepts. Mathematical objects defined within one branch of mathematics very often resurface in completely unrelated mathematical theories. This is such a common theme nowadays that we take it for granted. But the fact that we are accustomed to this process, the fact that we accept it, does not mean that we understand it.

Case in point: I remember, very vividly, the first time I learned that the derivative of Sin(x) is Cos(x). It was in high school, where, after a few initial lectures, about the secant formula and the formal definition of derivatives, the teacher wrote the following on the board:

$$\lim_{\delta \to 0} \frac{\mathrm{Sin}(x+\delta) - \mathrm{Sin}(x)}{\delta}.$$

To this very day I clearly remember thinking: Good luck with that. For who in his right mind would ever expect that the above monster would be solvable, let alone that it would yield a simple formula. And not only a simple formula, but the result is Cos(x).

The teacher, as expected, just brushed it off and continued with lecturing. As if nothing happened. I cannot really blame her, for she did what the vast majority of teachers do. She just plowed through the textbook. But I was stunned. Shocked actually. How could this be? There is absolutely nothing in the definition of Cos(x) that would even remotely connect it to any tangent lines, let alone to the tangent of its compadre Sin(x). And yet, here it was. As clear as a day.

So, I went home and started digging through books in a naïve attempt to decipher the mystery, to find the reason behind this bizarre connection. Soon I found the main culprit, the "thing" that made Cos(x) appear. It was the trig summation formula:

$$\mathrm{Sin}(\alpha+\beta) = \mathrm{Sin}(\alpha)\mathrm{Cos}(\beta) + \mathrm{Cos}(\alpha)\mathrm{Sin}(\beta).$$

The formula is obviously correct, but it did not really explain much. It did not help me understand the magic. So, I dug deeper and examined the proof of this trig formula. Needless to say, this failed miserably. The whole process helped me appreciate ancient geometry, for the proof is a visual masterpiece, convoluted yet elegant, flawless, a logical *tour de force*. But I found zero insights. The mystery remained unsolved. I still had no idea why the tangent line of Sin(x) would have a slope that equals Cos(x).

Many years later, I chased another rabbit down its hole. This time I wanted to "see" why π is in the equation of bell curve. In other words, I wanted to see the bond between the *coin-tossing problem* and the ratio of a circle's circumference and its diameter (that is, the π). These two have nothing in common, and yet the formula that explains the coin-toss problem has an obvious π in there.

So, I rolled up my sleeves, took on the de Moivre[3] formula and went hunting for π. I found it all right. It appears only once, and this time the culprit was Stirling's formula. So, now I had to chase another rabbit down a different hole. The proof for Stirling's formula is rather convoluted, but one can see that there are no πs in it, except at one

[3] The theorem is actually attributed to both de Moivre as well as Laplace.

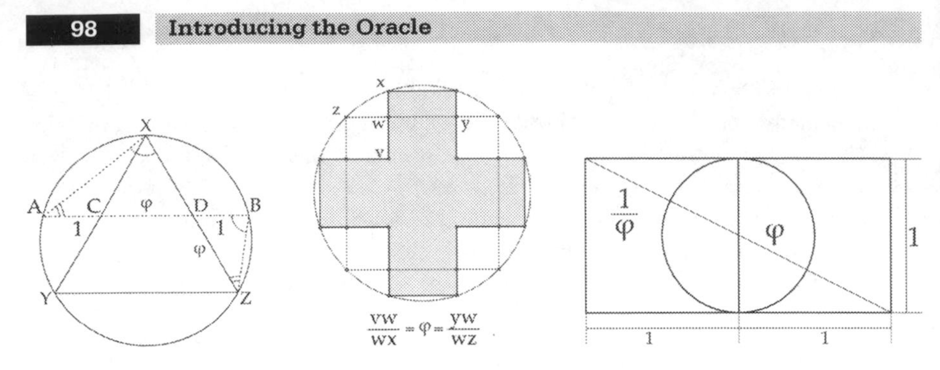

Figure 7.1 The golden ratio.

point: the limit of the (properly normalized) reminder term of the Euler–Maclaurin formula. But this, too, is not straightforward, for we need yet another formula, the Wallis product formula. And that is how we finally got π. And since we are here, we might as well state the Wallis product formula:

$$\frac{\pi}{2} = \lim_{n\to\infty}\left(\frac{2}{1}\frac{2}{3}\right)\left(\frac{4}{3}\frac{4}{5}\right)\left(\frac{6}{5}\frac{6}{7}\right)\cdots\left(\frac{2n}{2n-1}\frac{2n}{2n+1}\right).$$

Confused? You should be. Everybody should be. We chased the formulas and proofs like madmen, through different fields of mathematics, spanning centuries, and in the process we explained nothing. We are back to square one.[4]

Any mathematician skilled in the art can produce hundreds of similarly bizarre and unexpected connections. You can pretty much pick a topic or a formula at random, and a mathematician can create a whole network of distant objects that are somehow connected to it. But these connections do not seem to follow any logic or pattern. They were designed centuries, even millennia, apart, and for different purposes, and yet they seem to intermingle and connect with each other in such astonishing ways.

Just in a case there is any doubt about the inconceivable number of these rabbit holes, let us try an experiment. Let us pick a random, ancient concept, say the golden ratio. That is, take any positive numbers $0 < b < a$, for which the following is true:

$$\frac{a+b}{a} = \frac{a}{b}.$$

One can prove that this ratio must be unique (for any such *a*s and *b*s), and we will denote it with φ. The Greeks must have loved this number for it appears in so many so different geometric objects. Figure 7.1 gives a glimpse.

But, unexpectedly, and some 1,500 years later, we learned that this number not only relates to geometric figures but to integer numbers as well. In particular to the Fibonacci sequence:

$$1, 1, 2, 3, 5, 8, 13, 21, \ldots$$

[4] The expression "back to square one" comes from one of the three ancient geometric problems. It refers to our inability to square the circle.

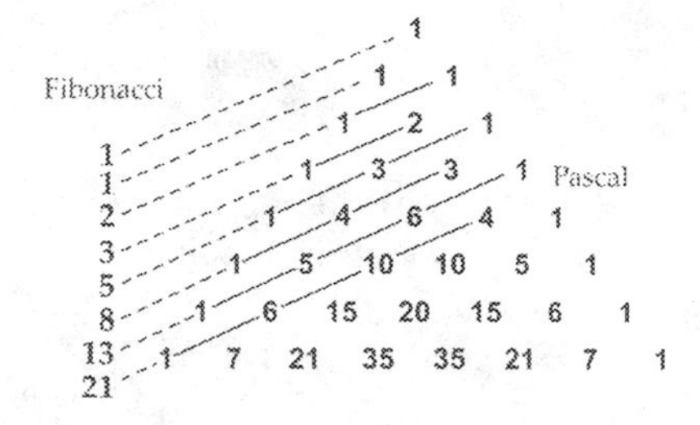

Figure 7.2 Connecting Fibonacci and Pascal.

where the next number is the sum of the previous two. This sequence grows very fast, and the 40th number in this sequence is 102334155. Now imagine trying to investigate how fast the nth Fibonacci number grows. Does it behave like 2^n or $n!$ or n^n? Well, it turns out that we can answer this question. This alone is rather surprising, but even more unexpected is the closed formula for the *nth* Fibonacci number. See for yourself:

$$f_n = \left(\varphi^n - \frac{(-1)^n}{\varphi^n}\right) / \sqrt{5}.$$

And, yes, this φ is the golden ratio. The formula is strange on its own, even if we do not factor in the mysterious appearance of the golden ratio. Namely, f_n is an integer (obviously) but the formula on the right has roots and irrational numbers. However, somewhat mysteriously, all decimals cancel out and we always get an integer, and always a Fibonacci integer.

Needless to say, the proof for the above formula reveals nothing. It requires some fairly sophisticated mathematics, recursive relations, and infinite series, and it all culminates with the quadratic equation $x^2 - x - 1 = 0$, which yields φ. As I said, we get nothing. No explanation for this mystery.

But now that we are connected to the Fibonacci sequence we can continue down the rabbit hole. There are so many pathways to take, so let us pick one at random: For example, one can easily connect the Fibonacci sequence to Pascal's triangle since its entries appear as the sums of the diagonals shown in Figure 7.2 (why they appear there nobody knows, but they do).

Pascal's triangle is another crossroads, one that leads to so many different pathways. Thus, yet again, let us pick one at random. Let us color all even numbers in Pascal's triangle with one color and all odd numbers with a different color. In the limit (as we consider bigger and bigger triangles) one gets the famous Sierpinski triangle shown in Figure 7.3.

And this triangle is odd indeed. It is too "hollow" to be a two-dimensional object and yet too "full" to be one-dimensional. Using topological methods, we can compute its fractal (Hausdorff) dimension as $\ln(3) / \ln(2) \approx 1.585$.

Consequently, in the "blink of an eye," we have connected an ancient Greek concept (the golden ratio) with the Renaissance work of Fibonacci, which is related to the work of Pascal, which then connects to the modern world of chaos theory and

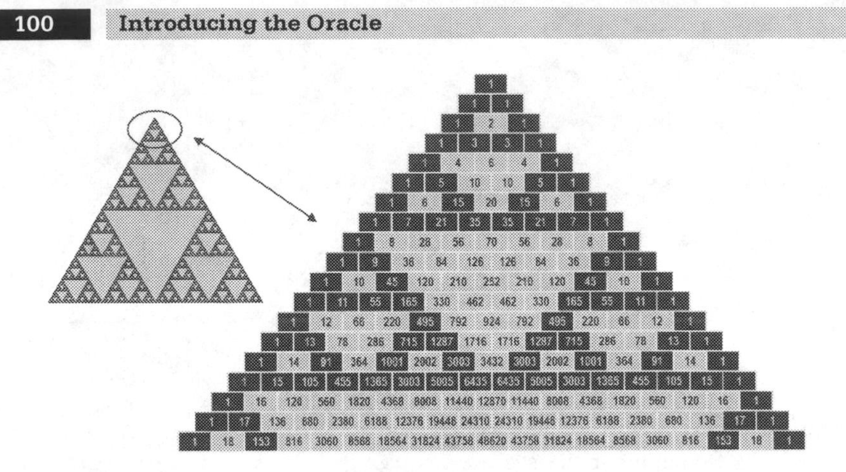

Figure 7.3 The Sierpinski triangle.

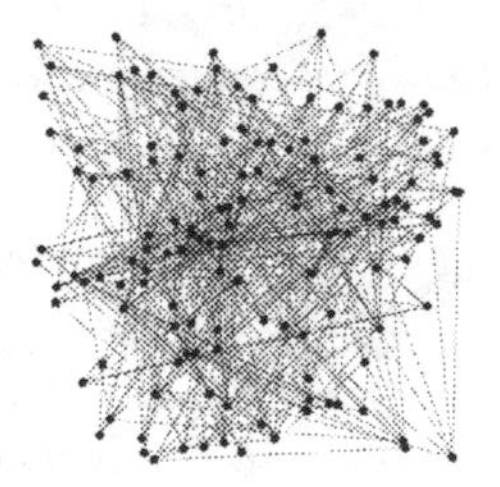

Figure 7.4 Mathematical connections.

topological dimensions. As I promised, in mathematics almost everything is connected to almost everything else.

But, as I also have promised, it all connects in haphazard, chaotic, and mysterious ways. We cannot easily detect any logic or reason, any big picture, any grand design. Thus, for better or worse, the most precise, the most logical of human activity, its majesty mathematics, is the one thing that in its essence offers a complete mess. It consists of thousands (if not millions) of theorems and concepts, too many of which are connected in some bizarre, incomprehensible fashion.

A picture is worth a thousand words, so here is an illustration. Imagine if one were to catalog and record these bizarre haphazard connections, all the pathways between different appearances of π, φ, Cos, Sin, log(log(n))'s, and such. If one were foolish enough to try this (and I must confess I did), one would get an incomprehensible jumbled and chaotic chart not unlike the one presented in Figure 7.4.

Nevertheless, this particular interconnectivity of mathematics, as mysterious and inauspicious as it might seem, is the very reason mathematics exists. Imagine the alternative. Imagine a situation where we have distinct mathematical areas, like islands in the Pacific, standing desolate and alone. And why not? It is conceivable, even expected, that people studying the coin-tossing problem would be on their island and completely isolated from number theorists and their prime numbers. There is nothing that even remotely connects the rolling of a die with integer numbers with no divisors (that is, the primes). And to be fair, for some 2,000 years, these two concepts, as well as many other mathematical branches, did remain isolated from each other.

Which is to be expected, and which is what happened with other human activities. For obvious reasons, we have the following pattern: Distinct human activities yield distinct theories and narratives. Nobody expects to see interplay between the work of a psychologist and a geologist. No sane person would imagine any deep connection between botanists and economists. And yet, in mathematics we see the exact opposite trend. We see profound connections between some fundamentally different concepts. For example, the axioms on which the "game of chance" is founded are identical to the ones we use to define area and volume. I am sure Euclid would have been bewildered.

Well, bewildered or not, we must admit that distant mathematical concepts do interplay, and they do so in spectacular ways. Ever since the ancient Greeks entered this amazing maze, this splendid majestic jumble we call mathematics, we have been searching for, discovering, and exploring new pathways, new connections, and hidden tunnels. Yes, it all looks chaotic and incomprehensible, but in fact it is not. There is a grand design, or at least we are starting to see its contours. And this underlying structure, this well-organized pathway of mathematics, is our oracle. One that guides mathematicians to do the right thing. As we shall see shortly.

8 On Probability

It might come as a surprise to some, but in many ways the branch of mathematics that deals with probability is the one whose makeup most closely resembles physics. While the centerpiece of physics (or any science) is the experiment, the same can be said for the probability theory; its very existence is based on a random experiment (a coin toss or a roll of die). Similarly, we can see that theoretical physicists provide a mathematical foundation for experimental physicists, while probabilists do the same for statisticians. In what follows we will examine this world in some detail, but more importantly, while doing so we will uncover the work of the mathematical oracle. We will see, firsthand, how it imposes itself and how it guides us through this infinite maze of mathematics.

8.1 Poisson Process

Probability theory is a branch of mathematics that studies randomness. The very fact that there exists a whole branch of mathematics that studies something *made without definite aim, purpose, or reason* (Webster's Dictionary) is surprising. But even there, in the realm of uncertainty and randomness, mathematics manages to impose itself. It turns out that this unusual setting of chaos and randomness provides us with very illustrative examples of this bonding between math and physics.

Arrival process. One of the most commonly observed random process is also one of the most fundamental. Sitting in a coffee shop or a fast-food restaurant we can observe arriving customers. If we do so, the first thing we notice is the haphazard and unpredictable nature of this procession. How many times have we experienced a long waiting line in front us, only to realize, as soon as we order our coffee, that the line behind us has cleared? Customers enter at a random, arbitrary rate. Often there is a cluster of people arriving, followed by an idle period with no one in sight, and then a long, steady, "even flow" of arrivals. An attempt to explain this chaotic parade seems doomed to failure. *What can one say about it? It is random.*

And to make things worse, these types of situations, these types of processes do not apply only to coffee shops alone. They are everywhere. The arrival of patients at a hospital or customers at a bank, the frequency of insurance claims, car accidents, telephone calls, website hits, plane crashes, hurricanes, etc. Each of these examples presents a unique case and each one is haphazard, random, and unpredictable. Thus, trying to address these types of different and chaotic situations, and trying to explain and classify each one with unified mathematical theory, well, that just seems ludicrous.

But, let us try, nevertheless. First, we should observe that our task resembles Newton's, for he, too, faced a chaotic and unpredictable world of motion and he tried to describe it mathematically. Newton's first step was to identify the key variables. In his case these were *force, mass,* and *acceleration.* We will do the same. In our case the essential variables of interest are the *number* of arrivals within certain interval and the *length* of time between two arrivals. Not surprisingly, just like in Newton's case, our variables are intrinsically related.

Newton simplified the problem and assumed a frictionless idealized case first, and so will we. We will assume that the nature of these arrivals does not change with time. An example will help: Suppose we observe a coffee shop during the time interval $[S, T]$, where $S = 2$ pm and $T = 4$ pm. It is safe to assume that this interval will provide a steady stream of custumer arrivals that is not influenced by outside factors like lunch breaks or rush hours. In other words, if we are able to mathematically model this (somewhat) idealized case, we should be able to enhance it to address the additional impurities (daily cycles, etc).

Note from the author: If some notation and discussion looks intimidating and puzzling, feel free to skip a few lines here and there.

Next, we work on quantification. Let us fix any subinterval $[a,b]$ (within 2 pm to 4 pm) and denote by

$$S_{a,b} = \textit{the number of arrivals during the period } [a,b].$$

Example: $S_{1,2}$= number of custumers arriving between 1 pm and 2 pm.

What can be said about this number $S_{a,b}$? Clearly, $S_{a,b}$ can only take integer values (since we cannot have 1.43 custumers arriving), and consequently,

$$S_{a,b} \textit{ could be } 0 \textit{ or } 1 \textit{ or } 2 \textit{ or } 3 \ldots.$$

Moreover, $S_{a,b}$ is obviously random, so it would make sense to ask the questions like these:

What is the probability that $S_{a,b}=0$?
What is the probability that $S_{a,b}=1$?
What is the probability that $S_{a,b}=k$?

We simplify the notation with: $P_k = P(S_{a,b} = k)$. That is, P_k is the probability that $S_{a,b} = k$. Example: $P_7 = P(S_{1,2} = 7)$ is the probability of seven customers arriving between 1 and 2 pm.

With this little digression we have arrived at the following characterization (description) of our random number $S_{a,b}$. Formally, we call $S_{a,b}$ a random variable and the associated table its distribution (formally: a distribution of a random variable $S_{a,b}$):

Possible outcomes	**0**	**1**	**2**	**3**	**4**	...	....
Probabilities	P_0	P_1	P_2	P_3	P_4	...	....

This seems like a good place to pause for a second. It might not look like much but we have just accomplished a major task. Although we started with a chaotic and seemingly incomprehensible problem, we managed to put it into a reasonable framework. We now know what needs to be quantified: We need those P_k's. We might not know how to find the formulas for them, but at least we know what we should look for, which is a much better position than our utter confusion of just a few minutes ago.

But we must be fair and admit that our position is still rather desperate. We have no idea how to determine any of these P_k's. And there are infinitely many of them. Can P_1 be $\frac{1}{2}$ or $\frac{1}{3}$? Is P_0 always bigger than P_5, is it true that P_k's always decrease (i.e., $P_k > P_{k+1}$)? So many questions. And we have no way of answering any of them. So, we go back to Newton, and we try with postulates (axioms). We will characterize (formalize), the essential ideas or properties that this process (or this variable $S_{a,b}$) should exhibit.

Axiom 1. (Stationarity) For any h, the probability distribution of $S_{a,b}$ and $S_{a+h,b+h}$ are identical.

In simple terms: If I count the customers arriving between 2:00 and 2:30 pm while the reader does the same but for the shifted interval 2:10 to 2:40 pm, we should have identical chances of seeing, say, five customers. That is, both intervals are of the same length (30 minutes), and the fact that I started earlier should not matter.

Axiom 2. (Independence) For any disjoint intervals [a,b] and [c,d], the random numbers $S_{a,b}$ and $S_{c,d}$ are independent.

In simple terms: If the reader is counting customers arriving between 2:10 and 2:40 pm, the fact that some customers had arrived earlier (prior to 2:10 pm) cannot influence the reader's total count.

That's it! Just two simple and very intuitive assumptions design to capture our impressions of this haphazard process. People are entering, completely randomly, and the rates of these arrivals cannot be influenced by what happened earlier (Axiom 2 - independence). This haphazard pattern does not change in time; it stays as it is (Axiom 1 - stationarity). Keep in mind, at this moment we have no idea what that "haphazard pattern" is, we only assume that it does not change with time.

There are two more axioms, but they only address some technical issues, and they are very reasonable. One deals with the fact that we should not allow for simultaneous entrances (it make sense, we "forbid" two people from entering at exact same time), while the other axiom deals with extremely long and extremely short intervals. If we wait *long enough* someone must enter the shop, and no matter how *short the interval,* there is always a chance that someone might enter (both extremes make sense).

The essence of the process (our experience) is captured by the first two axioms, while the last two, although perfectly acceptable, are there mainly to ensure mathematical rigor. I hope the reader will agree that the above four rules are very simple and intuitive. In many ways we are in a position similar to the one Newton faced: We have a few very simple and intuitive rules and we will try to build a theory out of them. That was the similarity, and here comes the difference: Newton assumed that *the larger the force the larger the acceleration,* which was reasonable, but he went one step further and declared the formula: the celebrated $F = am$. Which was a gigantic leap of faith. He was right (until relativity came along), and also very lucky.

Well, we did not make any leap of faith. After all, we are mathematicians, thus a bit more conservative than physicists. We did not impose any formulas nor equations regarding our P_k's. Frankly, we do not know if formulas for P_k's even exist, let alone what the correct equations would be. All we did was to assume some general and very reasonable behavior within this process. The obvious question presents itself: *What next?*

And here I would ask for some patience and some humility, for very soon the oracle will show its face, and we will have to be respectful. And proceed with care. Let us reflect on our position. Now that we have our rules (and no equations whatsoever), there are a few scenarios.

First scenario: Too many solutions. Since we did not impose many restrictions on our process and no direct requirements regarding our objects of interest (that is, no formulas for the P_k's), it is more than likely that there are just too many possible solutions. In other words, our axioms might not be restrictive enough. A calculus example will help: Imagine that we have described a certain physical phenomenon via a function, and that we have requested that such a function must be *differentiable, periodic,* and *positive.* You see the problem? These restrictions are too easy to satisfy, and we could construct infinitely many fundamentally different solutions. That is, our mathematical description of this physical process is not of much help.

Second scenario: No solutions. Again, this is a perfectly realistic possibility, and we have seen it happened before. Remember the case with measure theory, where we imposed some very reasonable requirements for a "measure function" only to show that no such function exists. The rules were too restrictive. A calculus example: Imagine the restrictions calling for functions that are *differentiable, periodic,* and have *finitely many zeros.* It is easy to see that such functions do not exist (one zero – due to periodicity – implies infinitely many zeros) and these rules would be too restrictive.

Tricky, isn't it? Unlike Newton, we did not impose our own formula. We took the "high road," but now we are paying the price. We have this predicament. Both of the above scenarios are troublesome, and both would require additional modifications. And, let us not forget, even if we resolve this issue, we are still "miles away" from the solution. We are not even sure if it is possible to derive the desired formulas for P_k's, let alone to how to construct them.

But what if, by some miracle, we can show that a solution exists and that there is only one such solution? In other words, imagine that we can show that there exists only one process that could satisfy the above four axioms. Wouldn't that be great? And, since we are dreaming, why not go for broke and fantasize that we can actually express this solution via a simple and elegant formula?

And, what do you know, we got lucky. One can indeed prove that there exists *one and only one* (up to a constant λ) stochastic process that would satisfy the above four axioms. As if this was not enough, we can mathematically describe these solutions with very simple and elegant formulas. Judge for yourself:

$$P_k = \frac{\lambda^k e^{-\lambda}}{k!}, \text{ and } P(T > t) = e^{-t/\lambda}.$$

The second formula deals with the waiting time T (time between two arrivals).

The above formulas describe the celebrated Poisson process. This is a well-known mathematical construction, and Poisson's formula has found its use in so many so different places; in pure mathematics but also in engineering and science. Interestingly, this celebrity status, this wide acceptance of the formula, is the main reason for our ignorance. Namely, we are so used to the Poisson process that we often forget how lucky we got. To be able to capture so many, so different, and so random and haphazard phenomena, with such simple equations, well that is nothing short of astounding.

The proof of this result is not too long (a few pages) but it is not too simple either. It requires some knowledge of characteristic and moment-generating functions as well as some differential equations (that is how we got e^x in there). But, as is expected by now, the proof reveals nothing about the "big picture," nothing about the "meaning" of all this. One might wonder: How did this happen? Why did we get so lucky? But the proof says: It happened because it happened.

The reader might be confused about all this, but, I hope, she is also aware that something profound indeed happened. We can sense it. Just a few pages earlier, we entered the infinite labyrinth of mathematics and we faced an impossible task. We attempted to explain this haphazard world of random arrivals, a world with so many so different scenarios, all chaotic and intractable. And although our task seemed hopeless, once we started, we solved it fairly quickly. Not just one of these processes – *we solved them all.*

As if by some kind of divine intervention, we walked through this infinite maze intact. The oracle of mathematics cleared the way for us. It forced all of those different and chaotic processes to behave the same way. Clearly, hurricanes and people entering a coffee shop are different, and so are the hacker attacks on our website, and yet, the oracle says: *No! They might look different to you, but their behavior is the same and there is only one way to mathematically describe them.* And, since the oracle is kind to us, it offered the ultimate gift: *simple formulas.* And for that we should be eternally grateful.

8.1.1 Since We Are on This Topic ...

Imagine the peak of hurricane season in September 2200, some 200 years from now. Clearly, we have no idea how many hurricanes will make a landfall on the continental United States during that period, but we do know a few things. The first week of this month has the same chances of seeing hurricanes as any other week (Axiom 1), and whatever happened in the first week will not make any difference in the weeks that follow (Axiom 2). Since we measure time very precisely no two hurricanes will arrive at the same microsecond, and if we pick any period of time, say an hour, there is some chance that a hurricane might make landfall within that hour (some 200 years from now). Thus, all four axioms are satisfied and oracle says: *This arrival process must be a Poisson process.*

------------------<>------------------

We finish with an anecdote. Many years ago, I worked with engineers in an industrial lab,[1] helping with the algorithms for elevator calls. In a building with more than one elevator, the current algorithms assign the closest elevator that is going in your direction (to pick you up). This is clearly the simplest algorithm to implement, but also, clearly, it is not the most efficient one. Our team tried to design a better method. The tricky part was the obvious random nature of customer arrivals. It did not take long before the Poisson process presented itself. I had to ask:

> There are so many other stochastic processes to consider. Are you sure the arrival is based on a Poisson process?
> Ooh, yes, definitely Poisson. That is a well-known fact.

Now, I was the only mathematician in this group, and the engineers, I have to admit, were very tolerant, and even appreciative, of my annoying questions. After all, their designs were going to be implemented in real-life machines, and naturally, they would rather discover a mistake earlier than later. So, in this spirit, I continued:

> Well, if it is indeed well known, can somebody explain to me: Why Poisson?

Silence. Nobody answered and they just looked around in a futile attempt to find a person willing to justify this "well-known" fact. Then Dave, the group leader said: *Go, fetch Alex.*

Alex was a Russian mathematician, a few offices down the hall. He once told me: "I had an advanced degree and worked for the Soviet military. Americans gave me the Green Card in no time." So Alex showed up, we told him what we needed, and he took a marker and started scribbling on the whiteboard. Some ten minutes and a few whiteboards later, we had our answer. And it very much resembles the previous few pages on the Poisson process.

8.2 Wiener Process

In this spirit, let us now revisit Section 2.4 on flowers and very spiky curves, and the Brownian motion that was introduced there. As we saw, it all started with botanist Robert Brown, who observed the erratic zigzag movement of pollen particles.

[1] The lab in question is the United Technologies Research Center. This was the place where I stumbled on the "bouncing balls universe paradox" and the place where I started working on this book – some of the best years of my life.

Similar erratic and unpredictable behavior was observed by Bachelier, but instead of pollen he was tracing the fluctuations of the stock market. The final musketeer in this tale is Einstein, who noticed that the chaotic path of molecules exhibits very similar, strange, and unpredictable fluctuations.

Although the observed behavior was similar, the three "motions" were produced by three very different mechanisms existing in different dimensions. Stock prices are one-dimensional (they fluctuate up and down), while the movement of pollen particles is observed on the surface of a liquid, and it is two-dimensional. Molecules, obviously, move in three dimensions. Consequently, we are in a similar position to before, with the arrival of customers. We have some intuition, we have some observations, and we have no idea how to proceed. It seems impossible to mathematically model these types of erratic, haphazard, and unpredictable phenomena.

Before we embark with mathematics, let us play with physics. In a Newtonian spirit, let us start with some basic, fundamental properties. We will try to characterize the observed attributes of these strange and unpredictable "motions." We mark the position of a particle (or price or molecule) at time t as B_t, and we list some basic, empirically observed, postulates for B_t:

Postulate 1. The Translation and time invariant,
Postulate 2. The Scale invariant,
Postulate 3. The Erratic and unpredictable.

In plain terms, postulate P1 states that one is free to observe a particle at any time or at any particular position. In each case one should experience the same phenomenon: chaotic and erratic movement in a random direction. Postulate P2 states that this motion, observed at different time intervals (say every 1 second or 1 minute or 1 hour), if properly rescaled, behaves the same. Finally, the last postulate addresses the obvious observation regarding the unpredictability of the particle's path. In other words, whenever you start, or whatever time interval you use to observe this motion, it always behaves the same: unpredictable, erratic, chaotic.

Pictures are worth a thousand words. Figure 8.1 shows examples of typical paths of pollen particles under different resolutions and time intervals.

Although these postulates might make sense to a physicist, they do very little for mathematicians. We are still utterly confused and not really sure how to proceed. A little help from our oracle would be welcome. However, the oracle, although kind and generous, is also very strict and stern. It will give us the answer, and it will hold our hand and walk us through the infinite maze of mathematics, but not before we ask the right question. Just like its cousin in Delphi, it remains silent, and waits for the proper query.

And to be fair, we do not even know how to pose our question yet. In the case of arrival (Poisson) processes, we managed, rather quickly, to quantify our problem and to identify what is needed. Remember those P_k's ? We needed their formulas. What about this case? What is the object, the quantity that we need here? Our quest is to somehow characterize a particle's movement in d-dimensional space. In

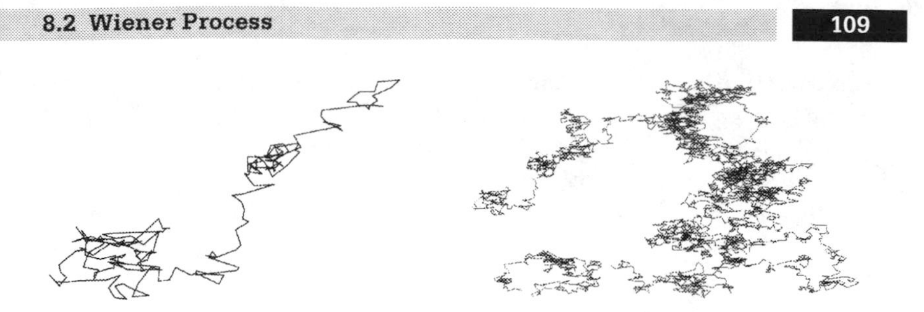

Figure 8.1 Brownian motion.

mathematics, this is done via parameterized curves: We let $X(t)$ represent a position of a particle at time t. Calculus textbooks offer a wealth of such curves to study. None of which have anything to do with our phenomenon.

Why? Because our curve is erratic, transition-and-scale invariant, and you will not find such a function in any calculus book. To make things worse, our motion is random as well. That is, the whole function is random. While we have learned how to model a random experiment (rolling a die) via random variables, here we have a completely different "beast." Each time we "roll a die," we get a whole new random function (new random path). How to even start addressing such a problem?

The mathematics behind these objects (random functions) is intricate (to say the least), and it took a while for us to design the appropriate apparatus. We had to learn how to deal with probability on infinite dimensional spaces, and how to extend the ideas of probability distributions into this infinite setting. And only then, only after we had learned how to properly ask the question, did the oracle respond. And what a magnificent response it was.

Query for the oracle. The proper mathematical way to treat these phenomena is to treat B_t as a stochastic process. The very concept of a *stochastic process* is a rather abstract one and even among mathematicians only a select few are fully familiar with it. Nevertheless, one should be able to follow the narrative of this tale without knowing all the details. It is enough to think of B_t as a random (zigzag) function that typically looks like one describe in the Figure 8.1. Without a loss of generality, we will assume that our process starts at the origin (thus $B_0 = 0$). We start with these axioms:[2]

Axiom 1. B_t *is a continuous function.*
Axiom 2. $E(B_t)=0$, *for all.*
Axiom 3. For all $0 \leq r \leq s \leq t \leq u$ ($B_u - B_t$ *and* $B_s - B_r$) *are independent (and if* $u-t = s-r$ *identically distributed as well).*

The first axiom ensures that the process models a path of a real particle and this path should not be discontinuous (for if it is, we would have a "particle" that can travel a

[2] Many textbooks include the Gaussian distribution assumption as the axiom. This is not necessary, for it follows from Central Limit Theorem. Finite variance assumption is often used, but it too is not necessary. It follows from the Continuity Axiom.

fixed distance in 0 seconds). The second axiom deals with the fact that there should not be any preferred direction. The symbol *E(.)* stands for *expected value.* In plain terms, this means that on average the process moves equally likely to the left or right, up or down.

The third axiom is a little bit more complex. This is where we mathematically address the most captivating property of Brownian motion: erratic, almost lifelike movement. The axiom states that the *increments* of this motion are independent of each other. In other words, if one observes a particle for a period of time, one cannot say anything about the direction (increment) of the particle's path for the next period of time; it could move in any direction. The increments of its future trajectory are completely independent from the past behavior.

The axioms do not impose any formulas but only some generalized descriptions, and it is possible (even likely) that we will end up with a conundrum. The situation is nearly identical to the one we faced with the arrival processes. We have the axioms, we have the phenomenon, and now we face a dilemma. There are two troublesome scenarios:

1. Too many solutions (too many stochastic processes satisfying the axioms).
2. No solution (there are no processes that satisfy the axioms).

Of course, there exists another (unlikely) possibility. What if we got lucky? What if there exists *one-and-only-one* solution? The reader is probably guessing that this is indeed the case, and she is probably guessing that on top of this (astonishingly) fortunate outcome, we have been blessed with a simple formula describing this one-and-only-one process.

And yes, it is true. The oracle, yet again, was kind to us. It gave us the answer (Norbert Wiener was the recipient), and it is unique; we do not have to worry about different formulas for different situations. Price fluctuations, molecules, pollen, they all follow the same process (up to a constant), and to address different dimensions we just create a few independent one-dimensional processes and plug them into the desired coordinate. But the oracle's generosity knows no bounds and it bequeathed us very simple formulas as well. For any (positive) fixed s and t we have:

$$B_t = N(0,t) \text{ and } Cov(B_s, B_t) = \min(s,t).$$

Here, $N(0,t)$ stands for a normal (Gaussian) random variable and Cov is covariance.

It is worth meditating on this for a moment. Yes, the reader (most likely) had guessed this fortunate (one-and-only-one) outcome. And yes, these types of "divine interventions" have been common throughout the last 2,500 years of our mathematical adventure. And yes, we have gotten used to them. But we should not take them for granted. As I said, it is worth pondering on this.

8.2.1 Since We Are on This Topic...

According to Newton's postulate on inertia, a body will remain still or move along a straight line at a constant speed if and only if the resultant force on this body is equal to zero. This is one of the most fundamental laws of physics. It is Newton's first postulate, it was Galileo's postulate and it is the essence of Einstein's first postulate. So, as the laws of physics go, this is a rather important one. Nevertheless, a simple argument, inspired by this chapter on Brownian motion, contradicts this fundamental principle. Allow me to explain.

Since the path of Brownian motion is a continuous function, according to the laws of mechanics one could postulate a particle that would follow its path. Since this path does not have any preferred direction (Axioms 2 and 3) the resultant force on this particle must be zero (otherwise we are in contradiction with Newton's second law). But the Brownian path is obviously not a straight line so we are in a contradiction with the Newton's first law. Cute isn't it?[3]

One solution to this paradox is to modify Newton's postulate on inertia and allow for a third possibility: *A particle could move along the path of Brownian* (i.e., Wiener) *motion* (see more in the Appendix).

8.3 The Bell Curve

> There must be something mysterious about the normal law [bell curve], since mathematicians think it is a law of nature whereas physicists are convinced it is a mathematical theorem.
>
> —Henri Poincaré[4]

The bell curve phenomenon has been observed for centuries, by scientists, engineers, sailors, businessman, psychologists, seismologists, you name it. If one were to plot a bar chart, comprising the IQ scores of the readers of this book, one would get a bell curve. If one did the same with the height of the readers, one would also get a bell curve, although, clearly, IQ and height have nothing to do with each other. In fact, we could take almost any biological trait, say number of readers' cousins or the readers' age, collect the data, make a bar chart, and *voilà*, the bell curve presents itself.

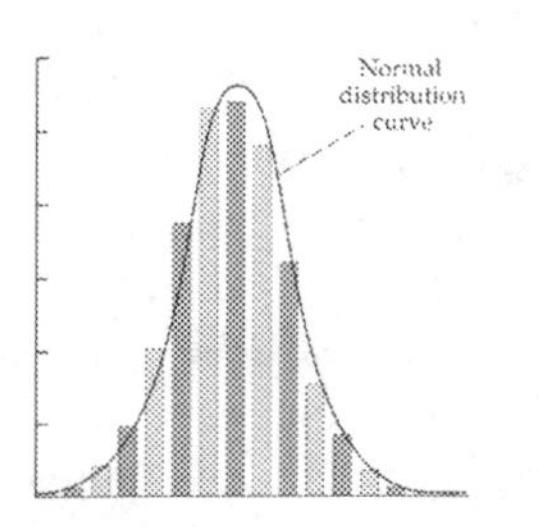

$$f(x)=\frac{1}{\sigma\sqrt{2\pi}}e^{\frac{-(x-\mu)^2}{2\sigma^2}}.$$

Figure 8.2 The bell curve.

[3] These are some ideas that Goran Peškir and I talked about one cold winter evening at Princeton, a long time ago.

[4] Mark Kac, *Statistical Independence in Probability, Analysis, and Number Theory* (he Mathematical Association of America, 1959), p. 52.

But why stop with biological traits? We could collect readings of odometers in readers' cars, and yes, that too would result with bell curve. Or something completely different: Take, for example, a bridge in some US state and inspect the percentage of rusted bolts. These numbers are obviously random – different bridges will have different percentages of rusted bolts. But if one collected such data for hundreds of bridges, and presented the findings as a bar chart, well yes, again, one would get a bell curve. If one counted 911 emergency calls per night in a large city, and then collected a year's worth of data, that too would yield a bell curve. It is everywhere.

And, to make things more peculiar, chances are that this omnipresence of the bell curve does not overly surprise the reader. Which only underscores my point: This curve has entered our lives so often and from so many different angles that we even forget to ask the obvious question: *Where does it come from?*

Think about it. We are talking about random measurements, collected from completely different sources, created by different mechanisms (metal oxidation, driving, genes, emergencies, etc.), and yet all this chaos and all these differences are somehow erased, nullified, as when we plot the data we always get the same thing. A miracle if ever there was one.

No wonder mathematicians and scientists were puzzled, and in turn knocked the ball into the other's court. For mathematicians, the whole process resembled a scientific endeavor (setting up the experiment, collecting data, plotting charts, etc.). On the other hand, the eventuality and the precision of the final outcome, the bell curve, clearly called for a mathematical result. So, this bell curve phenomenon, is it a law of nature or a mathematical theorem? Well, let us see.

The Central Limit Theorem. One thing is certain: The scientists who first observed this phenomenon had no idea how to address it. These occurrences resembled a law of nature, but this law was unlike any other. It was intractable, hard to "pin down." It had to do with randomness, and the formal, dictionary, definition of randomness is, *something without definite aim, direction, rule, or method*, so who in his right mind would study such a thing, one without any "aim or direction or method"? Naturally, scientists brushed it off as something for mathematicians to ponder.

Well, randomness is my bread and butter. I have spent a better part of my life studying this very bell curve (the Central Limit Theorem), so let me tell you a story. It is an interesting one in its own right, but more importantly, it fits perfectly with our mission. It illustrates the oracle in action, but this time we have the parallel process, since both scientists and mathematicians observed it simultaneously and they both tried to avoid it as long as possible. Probability theory, the branch of mathematics which studies randomness, started rather late, and rather slowly, for it is not easy to mathematically formalize something *without definite aim, direction, rule, or method.*

We begin our tale some 300 years ago, with Abraham de Moivre. He studied coins and observed that if one tossed a large number of them, say 20 coins, and if one added the number of heads, one would get a random number between 0 and

20 (which makes sense, it corresponds to smallest and largest number of heads among 20 coins). This part was obvious, but he went a step further and answered the following question: What is the probability that this sum is equal to a given number k? In more precise terms: If one tosses n coins, what would be probability that one would get exactly k heads? In particular, he asked what happens if n is large.

Well, as it turned out, this was not an easy task. The computations were cumbersome, technical, and quite lengthy. Moreover, the final result, the formula that tells us the probability that the number of heads is k (for large n), is far from an elegant one. In fact it is rather messy. Judge for yourself:

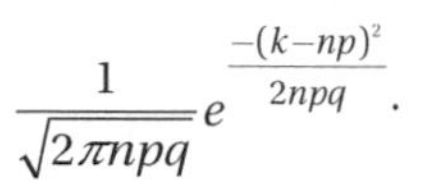

$$\frac{1}{\sqrt{2\pi npq}}e^{\frac{-(k-np)^2}{2npq}}.$$

So, why am I bothering you with de Moivre's tale? Because it produced the bell curve. A trained eye can spot it. It comes from the form $e^{-(..)^2}$. (What is π doing there, well, nobody knows.)

This was an important result. Obviously. It also placed scientists, as well as mathematicians, in a tricky position. There were pros and there were also cons. True, this result was able to derive the bell curve mathematically, but the mathematics was not pretty. The computations, as well as the resulting "monster formula," were a far cry from the simple and elegant mathematical models we were accustomed to seeing. Moreover, the whole exercise involves one very specific, controlled type of randomness – "tossing a coin" – which is too simplistic and too removed from the real-life situations in which the bell curve was observed. These were the "cons."

Here are the "pros." We did get the right curve and we got it by mimicking the original premise: random experiments. Moreover, if one examines de Moivre's actual result one finds that it also covers "unfair" coins (those "p and q" correspond to probabilities for the "head" and "tail" of a coin), which, in turn, mathematically justifies the bell curve appearance in the examples of "rusted bolts" as well as the "911 calls." The model is as follows: The probability that a bolt is rusty is "p," thus, examining n-bolts on a bridge is similar to tossing n unfair coins and counting the heads (which is what de Moivre's formula does). Analogous reasoning applies to 911 calls (the probability that a person dials 911 is p, and n-people in city corresponds to n-coin tosses).

So, it does work! We got the desired mathematical explanation, one that seems to cover some fabricated examples (coin tosses) as well as some real-life scenarios. But there are many more "cons" to come. De Moivre's computations are very sensitive and they work only if coins are *identical* and *independent,* something that is nearly impossible to reproduce in nature. It is doubtful that all the bolts on the bridge would have the same likelihood of being rusty (the "coins" are not identical). Moreover, rusting is a localized process, thus an observed rusty bolt is probably surrounded by other rusty bolts (the "coins" are not independent). An analogous critique could be constructed for 911 calls (different areas of a city and different times of day will influence the probability of 911 calls).

But this is not all. We have a bigger problem to address, one that is persistent, regardless of whether we could or could not fix the above criticism. Namely, there are so many so different scenarios: *IQ scores, height, weight, car mileage, temperatures, blood pressures, cholesterol levels, earthquake strengths, hurricane wind speeds, tornado lifespans*–an endless list of examples. This list cannot be linked to any type of coin-toss scenario, and yet it keeps producing the bell curve. Take a person's *height* for example: Nutrition, genes, environment, emotional stress, political situations, and numerous other factors influence a person's growth. These factors are clearly random in nature, and clearly not related to a coin toss. And yet, the final result, height, when plotted, produces the bell curve.

We should pause here for a second, for this is an important milestone. So far, we have witnessed the work of the oracle, but only after we learned to ask the right question. Here, the process is reversed. We can see the oracle's work beforehand. Somehow, it seems, all the randomness coming from individual factors evaporates, and only the bell curve remains, as if some kind of mechanism, some kind of force, is pushing all these unrelated chaotic factors toward a common goal. We can almost sense the oracle pulling its strings. And this particular process, this "force," is unlike any other mathematical model we have encountered so far. Allow me to explain.

Ever since Newton introduced his three postulates, scientists have accepted the following unwritten credo: *The further from an idealized mathematical situation our model is, the more inaccurate the mathematical prediction.* This makes sense, and this has been observed countless times. A feather and a cannon ball dropped from the Tower of Pisa will not hit the ground simultaneously (despite Newton's and Galileo's insistence), not because the great masters were wrong, but because the real-life experiment does not mimic the ideal, mathematical model. The laws of motion are placed in an idealized, frictionless universe, in which all objects have a point mass and there is no air resistance. Reality is different, and the further we are from the ideal situation, the more "noise" we encounter and the worse our approximation gets. This is true with all mathematical models of the physical universe. Except with the bell curve.

It seems that the further we get from the idealized situation (tossed coins) the better approximation of a bell curve we get. Figure 8.3 shows a "height curve," and one cannot but marvel how well it is approximated by a bell curve.

The data came from thousands of young men who were called for military service in 1939. This population was very diverse with different ethnicities (Scots, Irish, English, Welsh), different social statuses (peasants, workers, miners, middle-class men, nobles), different access to food and health care, and all these influential factors were random and unpredictable. Thus, it would be reasonable to expect some random and chaotic chart. But no! We got a near perfect bell curve, the same one we got after tossing bunch of coins. And one cannot but wonder: *How come?*

So, we mathematicians had a "hot potato" on our hands. We solved it alright, but it was a bumpy road, and a long one. It was not until 1933 that Kolmogorov axiomatized probability theory (about two centuries after de Moivre's Theorem), and

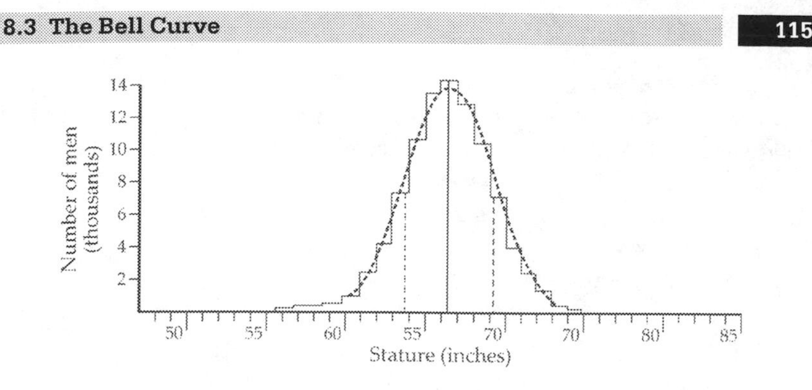

Figure 8.3 Heights of young Britons.[5]

the final verdict, the generalized result on the bell curve, was delivered just a few decades ago. It was indeed a rough ride, and not for lack of help. The mathematical heavyweights Gauss, Laplace, Bernoulli, Poisson, and so on paved the road.

We now have a full understanding of the mathematics involved. The bottom line is this: Yes, there is the mechanism, the "force" if you will, that takes all these random, chaotic, unrelated factors, crushes them, and makes the bell curve appear. The theorem is called the Central Limit Theorem, and, in its most general version, it states: If one takes many random measurements, and these measurements need not be identical nor independent, then, if one combines these measurements into one (by adding and normalizing them), the following will happen:

a. either this combined entity will converge,
b. or it will not converge.

So far this seems a bit dull. But here comes the punch line: If this quantity converges, it must converge to a bell curve, or, more precisely, to a Gaussian random variable.[6]

And this dear reader, explains everything. In layman's terms, the result states that if we add numerous random factors, regardless of their structure or interrelationship, we will always get a bell curve (or we get nothing tractable). This is why we see the bell curve so often and in so many different places. We mentioned *height* and the bell chart earlier. Here is the explanation: Dozens of different genes, numerous environmental and sociological factors, are random, and they all either inhibit or encourage growth. The final product, a person's height, will depend on the cumulative effect of these random factors. The conclusion (and the appearance of the bell curve) now follows easily from the Central Limit Theorem.

To reveal this work of our oracle, I like to play the following game with my students. I ask them to add their age, the last digit of their student ID number, their GPA, the number of cousins they have, the age of their car, the last digit of their phone number, and few more unrelated random quantities. Once they add all these

[5] Status of Britons called for military service in 1939. From Harrison et al., *Human Biology: An Introduction to Human Evolution, and Growth* (Clarendon Press, 1964).

[6] Technically no. If we allow for infinite variance then we can get different limits. Nevertheless, the vast majority of real-life applications have finite variance.

ridiculous and haphazardly chosen numbers (and the more the merrier), each student ends up with one specific number. I collect these, and then construct a bar chart. Incredibly, without exception, I always get a bell curve.

And that was our oracle. The theorem itself is its "gift" to us. It forces biology and physics and psychology and genetics. It forces the very creator of this universe to behave this way, and only this way. This "present" creates an order, a structure, within the infinite maze of mathematics, for it connects countless mathematical hallways together and channels them toward one common theme, one common highway, in this case the bell curve.

The Oracle, Its Majesty

Let us recall what we have just uncovered. We have demonstrated that mathematics is simultaneously hierarchical as well as haphazard. It is well structured and yet riddled with chaotic and inexplicable connections. Our dive down the rabbit hole (and pretty much the whole of Part I of this book) clearly demonstrated that mathematics is crammed with bizarre and unexpected connections, with countless formulas and concepts that interact with each other in the most inexplicable ways.

On the other hand, our discussion on probability theory confirmed what we have suspected all along: mathematics is a well-organized and hierarchical entity. From the beginning of our education, ever since the grade school, we have been exposed to this structure. We saw how nicely numbers and fractions interact, and as we learned more, we saw the coordinate systems and how geometry and trig functions worked together, and the more we learned the more of this orderly hierarchy we have uncovered.

If one climbs higher, even more of this structure becomes apparent. Limits, derivatives and integrals, are the next "layer," which then leads to differential equations, harmonic analysis, vector spaces, abstract algebra, and the list goes on. All these topics are just branches of one well organized tree-like structure. Figure 8.4 will help us appreciate this duality.

The chaotic chart on the left corresponds to protein-protein interactions in yeast and it very nicely depicts the idea of: Countless mathematical concepts that interact with each other in most unexpected ways. The orderly chart on the right represents the computer evolution, starting with 1946 ENIAC model and it nicely portrays mathematics as a well-organized and hierarchical entity.

The odd thing is that the grand design of mathematics, or at least our impression of it, is captured by both of these charts simultaneously. This strange and absurd impression, of mathematics being both chaotic as well as hierarchical, does not go

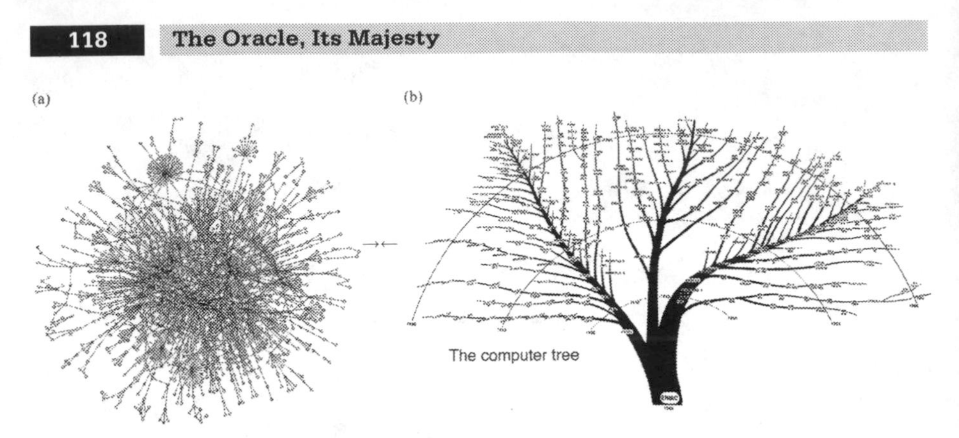

Figure 8.4 The convoluted and the orderly. (a) Protein-protein interaction in yeast. (b) Computer evolution chart.

away as one learns more. Not in the least. On the one hand, the higher one climbs the mathematical Mount Olympus, the more of the well-structured design one can see. But, then again, the higher one climbs, the more elaborate and twisted mathematical connections one encounters.

Take this gem: The formula $\log(\log(n))$ is a very (very) specific and unusual formula for which one cannot easily argue any scientific or engineering applicability. Nevertheless, this particular expression appears in mathematical statistics where it plays a prominent theoretical role (the law of large numbers). But of course, this is not the first appearance of this bizarre expression. Centuries earlier Euler showed that the sum of reciprocal primes is of the order $\log(\log(n))$. Nobody knows what connects the two.

Nevertheless, as the centuries roll over, and as we learn more and more, it seems that the second process, that of order and hierarchy, is the one that has the upper hand. Yes, as we progress, we encounter more of these bizarre and mysterious mathematical connections, but then again, as we march forward, we also manage to sort out the old ones. Of course, we still do not explain them (for who is to know "why" is derivative of $\mathrm{Sin}(x)$ equal to $\mathrm{Cos}(x)$), but we manage to catalog them. We put them in their place, connect them to other theories, and in the process we assemble the bits and pieces of this grand design of mathematics. The whole "assembly" process is an interesting one and it deserves a few lines.

Typically, after enough theorems (in a certain field) have been proven, and after enough "bizarre" and "unexplained" mathematical connections have been uncovered, a "prophet" shows up. In other words, a result is delivered, or a "sermon" if you will, that changes everything. We mortals, with our puny theorems and little tricks of the trade, we learn that all our hard work is just an echo, a reflection, of a far deeper and far more fundamental result. After this "gospel" is delivered, and after the initial shock absorbed, we quickly regroup, and like worker-bees we get busy. The theory is reassembled, simplified and cleaned. The books are rewritten. And now, with deeper understanding, and with even more vigor, we proceed with new theorems and new "bizarre" and "unexplained" mathematical connections, ... which then yields a new "prophet," and the cycle continues.

Astonishingly, this cycle remains unbroken, for some 3,000 years. We kept adding and adding, and at the same time, we kept rearranging and simplifying. In the process we have exposed a good portion of this grand design of mathematics, and, what presents itself is a fusion, a kind of symbiosis of the two charts presented earlier. Yes, we have an incredibly well-connected (and often haphazard) chart but at the same time, we can see the hierarchical, well-organized contours that very much resemble the computer tree diagram presented above. Of this hierarchical order we have already argued but a more skeptical reader is free to check the more detailed account in the Post Scriptum.

-----------------<>-----------------

The existence of this particular design, this haphazardly connected and yet well-ordered structure of mathematics, might be odd and unexpected, it might be bizarre and unfathomable, but it offers an elegant resolution to our question. Indeed:

The question: Why do we see ancient mathematical concepts finding their way into so many modern applications of science and technology?
The answer: Because in mathematics "everything" is connected to everything else, and in particular to ancient geometric concepts.

Let me walk you through it:

1. Pick any engineer or a scientist and observe that her work will inevitably involve some mathematics. Why? Because our universe is based on mathematical rules.
2. This, "some mathematics," will inevitably entangle itself with numerous other distant and unrelated mathematical concepts and theories. Why? Because mathematics is structured like that - it is incredibly interconnected (Figure 8.4a).
3. However, mathematics is also hierarchical. Consequently, any mathematician worth her salt, can now easily link our scientist with some ancient mathematical theory (Figure 8.4b).

The explanation is as obvious as it is simple. It "has to be" true. That said, one cannot help but notice that we explained nothing. We deciphered one mystery by invoking another.

Einstein asked: How can it be that mathematics, being after all a product of human thought independent of experience, is so admirably adapted to the objects of reality?

And we answered: Because mathematics has this mysterious structure.

Not much of an answer is it?

Epilogue: The Eternal Blueprint

We have come to the end of our journey. We have raised some important questions, and tried to offer some plausible answers. I will be the first to admit that it is easier to ask than to answer. Thus, the reader is free to take the suggested answers with a grain of salt. On the other hand, I trust that the reader will not do the same with the questions.

Through this journey we have learned a few things. It seems that mathematics forced itself onto, all of us, mathematicians and scientists alike. And, it seems, it did so onto the mighty creator as well. Try as you might to avoid it, mathematics, somehow, is always present. You study the stars and it is there. You study DNA and it is there. You randomly toss a coin and it is there. And it is the one that pulls the strings and tells us what to do and what not to do.

But, as mysterious as these mathematical intrusions might seem, it looks like there exists a grand design, a structure, a mechanism that pulls the strings behind the scene. We have exposed this mathematical mega-structure, this eternal blueprint that must be followed.[1] Everybody must follow it, the great architect and humans alike. And it is the existence of this structure that offers an elegant, almost trivial, explanation to our original question. Which, inevitably, yields to another, deeper question: *Why is mathematics structured this way?*

And this question, dear reader, I will not attempt to answer. This is the spot where I draw the line. I believe that the answer to the above question is: *Because it is.* It is the very same answer Moses got (*I am who I am*) and for a good reason. This is the question that touches the very essence of our universe, for this is the spot where the reader is as close to the great architect as she could ever be. No need for burning

[1] For more details on this structure see the Post Scriptum.

bushes or the parting of the Red Sea; we have a real miracle right in front of us. One could argue endlessly which of the holy books is the "right" one and which one was really written by the creator. But it would be hard to find a person that would disagree with the fact that the book of mathematics is written by the very creator; no matter who or what that is. And, like it or not, the universe is governed by the laws of mathematics.

There is one last story I would like to share. It is inspired by Carl Sagan's book *Contact*. In this book, the heroine, a young scientist, with help from an immensely advanced civilization, finds herself at the center of the galaxy. In the process, she finds out that this alien race has mastered all that can be mastered in the physical world: They can travel faster than light, they can travel through time, and they can leave their physical bodies. Manipulating the very laws of physics comes easy to them. Not only can they move their spaceships, they can also move stars, black holes, and rearrange galaxies. It seems that nothing in this universe challenges them. At the very end of her visit, the host asks our heroine if she has any questions, and she replies:

> Have you ever encountered anything in this universe that is above you, that you cannot explain?

The alien answers:

> Yes. We studied the digits of π, and after several trillion haphazard random digits we discovered that there exists a long sequence of digits that is made up of only "0's" and "1's." We know that this is mathematically impossible, and we believe that this is a message. What message we do not know, but it seems that the very creator of this universe had left this miraculous sequence for us to decode.[2]

I am pleased that Sagan places a simple mathematical concept above all the forces in the universe. It is π that puzzles the aliens and not the laws of physics. But there is more than meets the eye here and this tale deserves a few more lines. If there exists a message embedded in π, as this fictional work suggests, that would mean that the architect could manipulate the digits of π. Otherwise, this is not a message, right? But we have shown that π can be computed as an infinite sum

$$\pi = 4 - \frac{4}{3} + \frac{4}{5} - \frac{4}{7} + \frac{4}{9} - \frac{4}{11} + \ldots.$$

Therefore, the "message" depends on every one of these computations, since every single addition (or subtraction) will influence the trillionth digit. Thus, whoever embedded the message in the very digits of π would be the creator of *summation*. In other words, the oracle that made this message is the same that made: "1 + 1 = 2." And this ought to be some mighty powerful oracle.

Of course, this is a fictional story, but it relates to our quest in a profound way. We do not have this message embedded in π's digits, instead we have the message embedded in the very structure of mathematics. We have discovered this peculiar formation; this fantastically complicated and yet very hierarchical and orderly world. *The world that dictates the very essence of our universe*. Why is it there? What does it tell us?

[2] These lines follow the spirit if not the letter of the Sagan's book.

Only time will tell. Maybe. As for us dear reader, it is time to part. As I have mentioned many a time, it is much easier to ask the questions than to answer them. Hopefully you will agree that the questions raised here are important and interesting. I could not think of more appropriate way conclude but to use the following quotation:

> We have not succeeded in answering all our problems. The answers we have found only serve to raise a whole set of new questions. In some way, we feel we are as confused as ever, but we believe we are confused on a higher level and about more important things.[3]

[3] Posted outside the mathematics reading room, University of Tromsø, Norway.

Post Scriptum: On Mathematical Grand Design

The following pages are neither sufficient nor necessary for my argument (or this book), but I hope the reader will find them interesting. Namely, it appears that most of us, professional mathematicians or not, are very much aware of this grand design of mathematics, this mathematical blueprint that makes everything come together harmoniously. Except that we cannot easily find a visual representation.

Unlike biologists and linguists, who have created the complex yet hierarchical charts shown in Figure PS.1, documenting their tree of life (Figure PS.1a) and language tree (Figure PS.1b), mathematicians have done no such thing. Is it because we are "not there yet" or maybe because the bizarre interconnectivities of distant mathematical branches are obscuring the view? Or we just do not care? It is hard to tell.

Nevertheless, charted or not, this idea of well-designed and structured mathematics is commonly accepted. All one needs is to pick a well-developed mathematical theory (say analysis or algebra or point-set topology) and compare the standard textbooks. What one finds, even from random authors and from random countries, is that their tables of content are remarkably similar, and the early parts of these books are often nearly identical. Thus, there is a recognizable common ground, a

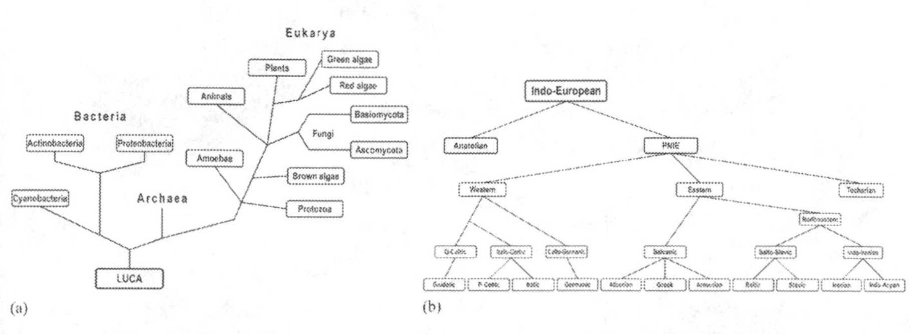

Figure PS.1 Grand designs. (a) The tree of life. (b) The language tree.

hierarchical order of things among many branches of mathematics. In what follows I will try to paint the big picture and offer a glimpse of the mathematical tree of life.

> A note from the author: In what follows we will address some nontrivial concepts and theories. The material is tailored toward the students of mathematics. Nevertheless, any reader brave enough is more than welcome to enter.

The Trunk

All mathematical theories start with axioms, a few and simple rules from which a theory arises. Clearly, one could construct myriad fundamentally different sets of axioms, and clearly each one would yield a different theory. And since in mathematics we have some very different branches that deal with some very different issues (numbers, the game of chance, logic, geometric figures, etc.), one would expect that these branches are based on some fundamentally different axioms. This makes perfect sense. Take the board games chess, Go and tic-tac-toe – they are very different games and they have very different axioms.

This seems reasonable and makes sense. The axioms behind the game of chance should be fundamentally different from those behind geometry. Well, reasonable or not, quite the opposite is true. In what follows I will show that many of the main branches of mathematics (themselves very different) have sprung from axioms that are surprisingly similar, nearly identical actually, and essentially all geometric. Allow me to demonstrate.

Geometry. This one is obvious. Euclid's Axioms are clearly geometric in nature.

Real numbers. We have already argued there is a very strong geometric connection here, but it pays to revisit this discussion. To start with, we must emphasize that the vast majority of applications of real numbers are *not about geometry* and that their origin is in *commerce.* We can confirm this with Mesopotamian artifacts as well as with modern examples. We use real numbers (i.e., decimals) to measure temperatures, pressures, force, and energy; to compute our mortgage payments, to pay taxes, and to balance our checkbooks. And yes, occasionally we use them in connection to geometry (length and areas typically).

Thus, we must conclude that geometry played a minor role when these numbers were first conceived, some 5,000 years ago, and plays an even smaller role now. Nevertheless, the axioms of real numbers are very much inspired by geometry. All but one of these axioms (on supremum) are deeply rooted in geometry. For example, the *commutative property:* $x + y = y + x$ claims that a length of two combined sticks (line segments) is equal, no matter from which side we start measuring (left or right). The *associative property* for multiplication $(xy)z = x(yz)$ is a declaration about a prism and its volume. The volume is defined as *base times height,* and thus the associative axiom states that prism's volume does not change if we place it on its side.

And let us not forget the disturbing *distributive property* (see Figure PS2):

$$ax+ay=a\,(x+y).$$

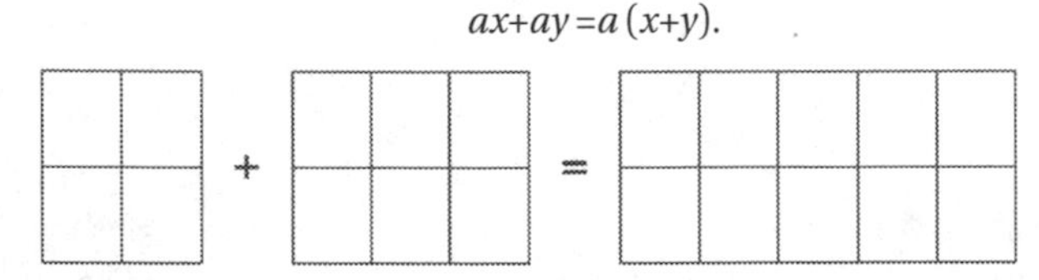

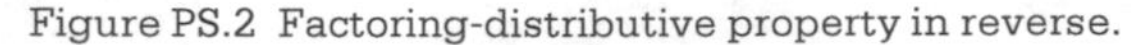
Figure PS.2 Factoring-distributive property in reverse.

Why this particular (geometric) property is disturbing will be demonstrated shortly.

Probability/measure theory. First, we observe that the axioms of probability theory are near-identical to the ones we use for measure theory (up to normalization). But if one examines the axioms of measure theory, one must conclude that they are about geometry. They specify a set of rules designed to precisely address the idea of an *area* or a *length* or a *volume*, which are obviously geometric attributes.

Indeed, the first axiom of measure theory states that the "measure" of a set is always positive (*geometric translation*: Length-area-volume of a geometric figure is always positive). The second axiom states that the "measure" of an empty set is zero (*geometric translation*: Length-area-volume of nothing is zero). The third axiom states that the "measure" of different objects, if combined, is equal to the sum of individual "measures" (*geometric translation*: Total length-area-volume of different geometric objects is equal to the sum of the individual lengths-areas-volumes).[1]

------------------<>------------------

A small pause is in order. We have just demonstrated that *real analysis, probability,* and *statistics,* as well as *measure theory,* are all based on axioms that are essentially just simple geometric rules. In fact, these axioms are so closely related to geometry that in all likelihood the ancient geometers used them as well. The Greeks dealt with areas, volumes, and lengths, and with near-certainty we can claim that they used rules that are identical to the ones described above, for real numbers and measure theory.

The above examples were about "concrete" objects, about numbers and the game of chance and such, so maybe if we venture into more abstract world the geometric connection will dissipate? Well, let us see.

Group theory. This theory serves as a foundation to one of the most theoretical branches of mathematics (*abstract algebra*) and, in many ways, *group theory* is a gateway to abstract mathematics. The objects and the operations used here are purposely left unspecified, generalized. Here we might say "u plus v," and we might even write "$u + v$," but we do not mean: $u + v$. In this world "plus" stands for some abstract operation that satisfies some abstract rules, and it is often denoted with different symbol, say $\oplus$. Here, mathematicians are completely disconnected from any physical manifestation (or interpretation) of the objects u, v, w, or the operation $\oplus$. All that matters is that the axioms are satisfied. Well, the time has come to see what these axioms are.

We start with a set A (any set) and denote its elements by u, v, w. Next, we define an operation $\oplus$ (any operation) and ask that it satisfy the following:

1. Closed operation: $u \oplus v$ must be an element of A.
2. Associative property: $(u \oplus v) \oplus w = u \oplus (v \oplus w)$.
3. Neutral element: There exists n such that $n \oplus u = u = u \oplus n$.
4. Inverse element: For every u there exists v such that $u \oplus v = n = v \oplus u$.

[1] The axiom allows for infinite number of these "different" objects. This notion is not foreign to ancient geometers but it is the main source for many "unusual" results and technical complications we see in measure theory.

It is astonishing how much mathematics follows from these few simple rules. They serve as a stepping stone to a profound and fascinating theory. However, we are more interested in the origin of these rules, and the reader cannot but notice that they are identical to the ones we used to axiomatize real numbers. Just replace the abstract objects *u, v, w,* and so on with regular numbers *a, b, c,* etc. Replace $\oplus$ with good old +, replace neutral element *n* with *zero* and the inverse element with *–a,* and *voilà,* we get the rules used for real numbers. And, as we have already demonstrated, they are geometric in their essence.

Vector spaces. Another interesting way to generalize the above ideas would be to fuse the two successful theories of *real numbers* and *group theory.* That is, we borrow the main objects from group theory. We keep them abstract (and unspecified) as before, and the only difference is that now we call them "vectors" (notation *X, Y, Z*). We also keep the main operation unspecified (abstract) as $X \oplus Y$. The only novelty is a new operation that fuses the abstract vectors with real numbers α, β, δ, etc. That is, we define a new (also unspecified) operation as $\alpha \circ X$. After this, we "agree" on a few rules (interactions between $\oplus$ and $\circ$) and we are good to go.

The theory which follows is nothing short of spectacular. The above abstraction allows us to see "the big picture," and discover a whole new world. It enables us to deal with spaces of infinite dimensions, to see structures and designs that would have been otherwise hidden. We will not discuss this particular topic (abstract vector spaces) in detail here. We are more interested in those "few extra rules" that we imposed. The ones that connect operations $\oplus$ and $\circ$.

Here they are:

Axiom 1. $\alpha \circ (X \oplus Y) = \alpha \circ X \oplus \alpha \circ Y$.
Axiom 2. $(\alpha + \beta) \circ X = \alpha \circ X \oplus \beta \circ Y$.
Axiom 3. $(\alpha\beta) \circ X = \alpha(\beta \circ X)$.

Needless to say, these rules look eerily similar to the rules we used to characterize the real numbers (just replace "$\oplus$" with the usual summation and "$\circ$" with multiplication). In particular, we see the *distributive property* (twice) for which we have already argued a very strong geometric connection.

------------------<>------------------

Another short pause is in order. We have just shown that *real numbers, probability and statistics, measure theory, group theory,* and *generalized vector spaces* are all based on axioms very much inspired by geometry. More importantly, these theories are responsible for the majority (perhaps even the vast majority) of mathematical intrusions into the worlds of science and technology.

As if this were not enough, we can do even more. Let us pick the most abstract of mathematical theories, one that deals with the pride and the joy of humanity: logic.

Mathematical logic. This is the branch of mathematics that is most strongly associated with the abstractions of the human mind, and one that should be the most independent of reality and experimental observations. However, even here we cannot escape the grip of geometry. Take one of the laws:

$$A \,\&\, (B.or.C) = (A \,\&\, B).or.(A \,\&\, C).$$

This statement of logic is nothing but a reflection, a mirror image, of a geometric rule. Just replace "or" with "plus" and "&" with "*times*" and we get our good old friend:

$$\textit{Distributive property } A(B + C) = AB + AC.$$

This particular observation is mindboggling and it deserves a few extra lines. Imagine that you are planning a day with your friend and have three things in mind: A = going to a movie, B = going to a restaurant, and C = going shopping. And suppose you tell your friend that you will: *(definitely) go to a movie* and *you can make a time for shopping* or *a restaurant (but not sure if you could do both)*. Formally, mathematically, you just stated: $A \,\&\, (B.or.C)$.

Now, if you would like to rephrase your statement, and break it into two, you would say: *Yes, I can go to a movie and restaurant,* or, *I can go to a movie and shopping (but not sure if I could do both)*. Formally, you just made the following logical statement: $(A \,\&\, B).or.(A \,\&\, C)$.

The obvious question presents itself: Why on earth would our logic, which after all is the very essence of our intelligence and humanity, be geometric? Well, nobody really knows, but apparently it is.

This said, I would like to emphasize that not all of mathematics can be easily reduced to a few statements about geometry. Clearly the Axiom of Choice is not geometric nor is the axiom on supremum for real numbers. Both of these created an avalanche of spectacular results, which (consequently) cannot be easily linked to elementary geometry. And there are others. The axioms of group theory are obviously geometric but they omit the commutative property (which is something geometry would insist on). Nevertheless, a surprisingly large number of mathematical axioms are rooted in geometry.[2]

------------------<>------------------

Thus, we can (safely) conclude that the trunk of our tree of mathematics should be rooted in geometry. And now that we have the trunk, we can start to build the tree. For example, it is commonly accepted that most of mathematics can be divided into two large branches, one of algebra and the other of analysis. So, we can plant these two branches on our tree trunk and build from there - by further subdividing into smaller branches, adding subfields, and sub-subfields of mathematics, thus creating the whole crown.

However, there is a problem. In typical trees the branches do not reconnect. In the tree of life, the mammals and the reptiles split, and we do not see them reconnecting ever again. In mathematics, on the other hand, we see the reconnections all the time: Two completely unrelated concepts, belonging to completely different branches, are often connected. Just remember all the rabbit holes we have encountered so far.

[2] For centuries, the words mathematics and geometry were synonymous. In fact, in the past, most mathematicians called themselves geometers.

Clearly, in the case of mathematics we have an additional structure. On top of the tree-like hierarchical design there is yet another layer, which is like a web, or a net, and which seems to interconnect everything. For in mathematics all concepts are somehow linked to all other concepts – and not in a hierarchical and organized way, but completely haphazardly. How to decipher this parallel structure? Frankly, I do not know, but I have some ideas.

Apparently, within mathematics we see certain themes. These are not necessarily bunches of mathematic or independent theories (at least not yet), but rather certain ideas, patterns that have been appearing through this symphony called mathematics. They emerge and reemerge in all unexpected places, always different and yet, under experts' eyes, clearly recognizable. They serve as glue, a connector, between all kind of different theories and branches. Maybe they could be a clue in our quest?

The Themes

The task of creating a blueprint for all of mathematics would be gargantuan, if not impossible. So, I must confess, in what follows I can only offer a few ideas, glimpses if you will, on how this tree of mathematics might be constructed. Oddly, I will borrow heavily from the natural sciences, biology in particular.

The scenario that unfolded with the natural sciences is well known. After enough specimens had been collected and catalogued, and after many a trained eye had scrutinized the data, the *grand design* of biology (or geology or linguistics), had presented itself. In the case of biology, the celebrated tree of life showed its face, and scientists could not help but notice that some common themes kept reemerging and reappearing with cunning regularity and at some unexpected places. Take the limbs of distinct species: the wings, the paws, the fins, the arms. They all look so utterly different and yet, once reduced to skeletons, a trained eye can spot the remarkable similarity. They all show the pattern so familiar to our own hand and arm. Strangely enough, similar eternal themes are present within mathematics as well. All we need is some patience and a trained eye to spot them.

Theme 1: Linearity

The idea of a straight line is as old as mathematics itself. But, interestingly enough, this concept, so dear to us humans, is ignored by nature. Try as you might, you will not easily find a straight line among the clouds or oceans, forests or deserts, mountains or rivers. Consequently, for more than 3,000 years (at least since the Egyptians), this concept was strictly geometric and used exclusively by engineers. This all changed with René Descartes and his "magic trick." His coordinate system opened the door for us and we learned that a straight line has its algebraic interpretation $Y = aX + b$. This changed everything.

After this "gospel" was received, slowly but steadily as mathematicians collected more specimens (i.e., proved more theorems) and after many a trained eye had scrutinized these results, the *grand design* presented itself. This theme of linearity is everywhere. And I do not mean this figuratively, but literally. A trained eye can spot the linearity within Newton's laws of motion ($F = am$ describes a linear relationship), in the axioms of real numbers (*Distribution Axiom* describes a linear relationship), and within the rules of formal logic. The crescendo comes with the *generalized vector spaces.* This theory is designed for one purpose only: to generalize the idea of *linear combination* (i.e., $\alpha X + \beta Y$).

Linearity is omnipresent within the natural sciences as well, whether biology or chemistry or physics. Take the following scenario which plays itself perpetually: One takes a "little bit of chemical X" (or force X or bacteria X), which one denotes by αX, and one takes a "little bit of chemical Y" (denoted by βY), and then one "asks" nature to combine them (i.e., conduct an experiment). And more often than not, much more often than not, nature combines them *linearly.* Yes, it is true: *When nature combines chemicals, or forces, or radiations, or waves, or germs, or enzymes it does so linearly, it computes, $\alpha X + \beta Y$.*

It would make sense to pause for a second here. We have just uncovered an important binding agent, or parts of a blueprint if you will. What do we have? On the one hand, we have an abstract mathematical theory that generalizes all possible linear relationships (general vector spaces) and, on the other hand, we have our universe, which apparently uses the very same linear operations.

Now any person worth his salt must conclude: Of course these two will interact. Of course many of the theories that mathematicians developed from these generalized linear interactions will be applicable in the world of nature which also acts linearly. How could they not? And although this is indeed true, and although many of the examples presented in this book are in one way or another a product of this particular theme, the action of this theme is not easy to spot – some training is required. Just like a biologist who needs some training (and a good microscope) before he concludes that all animal cells have nearly identical structures, we too must have a good eye (and a good "microscope") to properly detect the action of this linearity theme. A few examples will help.

Trig functions. At the first glance there is nothing "linear" in the appearance of wiggly charts of Sin(x) function. But a trained eye knows that appearance can be misleading. So, let us look a little closer.

One of the most fascinating of the *generalized vector spaces* are the spaces of functions. And among them, one particular space, the L_2 space deserves special attention. This vector space is not only related to but plays a fundamental role in so many very different applications of mathematics: signal processing, quantum mechanics, heat equations, Newton's mechanics, gravity, acoustics, to mention a few.

Now this vector space, the L_2 space, is of infinite dimensions, and as such, quite difficult to treat mathematically. As expected, there are many (many) more infinitely dimensional spaces we mathematicians encounter, but among all of them, the L_2 space is by far the easiest one to analyze, because for this space (and for this one only), mathematicians are able to construct a base. Thus, the only generalized, functional vector space that mathematicians can easily treat is the exact one that nature favors as well. What are the odds?

But there is more. Of course there is. The existence of the base is one of the key reasons behind the mathematical appeal of the L_2 space. Guess what the base is? Well, believe it or not, the trigonometric functions, good old Sin(x) and Cos(x), form the base. Trig functions are fundamental building blocks for this vector space that is favored by both mathematicians and nature itself. And the linearity theme has its fingerprints all over this concept: Any function that lives in this L_2 space (which is a generalized *linear* space) can be represented as a *linear* combination of trig functions.

Bell curve. Here comes another concept that seems as far removed from the linearity idea as one could imagine. The bell curve appears in relation to random, chaotic, haphazard things, so how could linearity intrude here? Well, let us see.

We have already commented on the *Central Limit Theorem*, the result of which forces so many very different random experiments to behave in very predictable ways. But there is more to this story. If one examines the theory, one can see that this result did not call for the bell curve. No, it states that any family of variables could be the limit, as long as the following holds: For any X and Y in this family, and for any numbers α and β, the quantity $\alpha X + \beta Y$ must stay within the family. In other words, the family must be impervious to the linear transformation. Thus, yet again, and quite unexpectedly, the *linearity theme* presents itself.

The bell curve (i.e., Gaussian random variable) is the only family of variables that satisfies this linearity property,[3] and that is why we see it in nature so frequently. Interestingly, another two families come close to this requirement: Poisson random variables are well adapted to summation ($X+Y$ is again Poisson), but not to multiplication by a number (αX is not Poisson anymore). Exponential random variable does the opposite. Indeed, αX is still exponentially distributed, but the sum is not ($X+Y$ is not exponential anymore). Thus, each of them satisfies "half" of the linearity property, and curiously, these two variables put together form the Poisson process, the very same mathematical construction that describes so many very different arrival processes.

Derivatives and integrals. Linearly combining the slopes of two tangent lines makes no geometric sense (What would be the meaning of *two slopes from the first line added to five slopes of another?*). Nevertheless, algebraically, differentiation and integration are perfectly suited to the *linearity theme.* Indeed, integrating $\int \alpha f + \beta g$ is easy, but if one tries the same for any other (nonlinear) combination (say, $\int f/g$, or $\int fg$, or $\int \sqrt{f+g}$, etc.) one faces a daunting, if not impossible, task. This is recognized by the theory and we view these two operations (derivative and integral) as *linear operators* on functional (linear) vector spaces.

Differential equations. On the one hand, we have scientists (physicists typically) who set up a differential equation that cleverly combines the fundamental laws of nature with the problem at hand. The solution of a differential equation would, in turn, describe a trajectory or a behavior of the physical system of interest. That is what physicists do. Mathematicians, on the other hand, are generalizing and studying all kinds of differential equations, regardless if they are of interest to the scientists

[3] As long as we assume that the variance is not infinity.

or not. They are deriving numerous tools necessary to solve as many differential equations as possible. So far, so good, but where is the linearity theme?

Well, it turns out that solving a differential equation is a daunting task and we humans are extremely limited in this regard. In fact, the only type of differential equations for which mathematicians have developed reasonably general tools are the *linear* differential equations (of the second order). This is very disappointing, since this type of equation presents only a tiny sliver of all possible equations to study. But, surprisingly, this great deficiency, this inability to solve the vast majority of equations, did not bother scientists that much. Why? Because quite often (surprisingly often) the equations physicists are interested in are indeed linear (and of second order). What are the odds?

To summarize: This simple theme, the linearity concept $\alpha X + \beta Y$, seems to be widespread through mathematics. I hope the reader agrees. We see it in axioms, we see it in abstract mathematical theories, and we see it in the applications of science and technology. Many a person (mathematicians in particular) would try to use this example as an "explanation." They might even claim that there is nothing to explain. *For didn't we just show how trig functions appear in so many fundamentally different applications? Didn't we just connect ancient geometry to signal processing and heat equation?*

Yes, we did, but we did not explain anything. A natural scientist of the eighteenth century could easily present the tree of life and connect different species, subspecies, genus, and families. He could do this in a great detail, but he could not explain anything. Darwin's theory of evolution did. Similarly, medieval astronomers could predict at what time of a year, at what hour of the night, Venus, or Jupiter, or Mars would appear. They could trace these planets through the night sky and they could do so with great accuracy. But they could explain nothing. Newton's theory of gravity did.

A mathematician can take trig functions from the ancient Greeks and, via L_2 spaces, connect them to the technology of signal processing or applications of heat equation or quantum mechanics. He could do this with great detail and precision, but could he explain anything? Could he produce a grand theory, the mechanism responsible for this structure? I do not think so.

But we digress. Here are two more themes that seem to play important roles within this mechanism, this tree of mathematics if you will.

Theme 2: Orthogonality

The story is near identical to one told earlier. We have a geometric concept studied by mathematicians but applied exclusively by engineers. For indeed, it is near impossible to find anything orthogonal built by Mother Nature. Consequently, for a good few thousand years the orthogonality idea was linked only to manmade objects or procedures. This all changed once mathematicians figured out how to generalize this idea, and then all of a sudden this concept proliferated through the world of mathematical theory as did its applications. We see it in physics, harmonic analysis, signal processing, image analysis, time series, regression, and so on.

Reasons, there are many, but it is safe to say that L_2 space played a major role. Namely, among a myriad of infinitely dimensional vector spaces, the L_2 space is the only one for which we can properly define orthogonality. This particular property makes it mathematically very appealing; in many ways it behaves like our usual three-dimensional vector space. But it was not only mathematicians who were impressed. The creator of this universe prefers it too; we can find L_2 space behind quantum mechanics, thermodynamics, and Newton's mechanics (among others).

Like many other tales in the book of mathematics, this one has a bizarre twist. Yes, we can define the orthogonal base (orthogonal vectors) in this space, but the question arises: What are these orthogonal building blocks on which the entire L_2 space is built? What are these foundations that yielded a vector space so fundamental to the world of science and technology? Well, believe it or not, trigonometric functions are these orthogonal building blocks. And please do not ask "why." They just are.

A few more details in support of the omnipresence of the orthogonality theme:

- The *orthogonality* of trig functions was the essence behind Fourier's solution of the heat equation.
- In many ways, the very existence of (as well as the name) harmonic analysis is a direct product of the *orthogonality* of trig functions.
- The cell phone revolution is very much rooted in this principle: The analog speech signal is processed by computing its *orthogonal* projection onto the linear combination of *Sin*s and *Cos*s.
- Regression analysis and the "fitted line chart" we so often see at business meetings, and in hospitals, chemistry labs, and city halls are nothing more than clever applications of *orthogonal* projection.
- The correlation coefficient, a number that had become so dear to so many users (and abusers) of data analysis, is very much rooted in the idea of *orthogonality* (i.e., zero correlation corresponds to orthogonal data).
- When de moivre worked on probabilities associated with a coin toss, little did he know that he too was dealing with *orthogonality*. Indeed, any sequence of independent random variables (independent coin tosses) is also an orthogonal sequence. The converse is not true, except in one (and only one) case: the bell-shaped curve.

There are many more applications of this principle within the world of technology as well as science. We see this idea proliferating through various mathematical theories too, and not only through analysis or general vector spaces. One can associate the idea of orthogonal complement with the ideas of direct complement or direct product seen in other, more abstract, branches of mathematics.

Theme 3: Circle

To be fair, this particular theme is more about the arithmetic operation of *squaring* (i.e., r^2) than it is about the actual circle. I chose this particular title since: (a) the formula for a circle is $x^2+y^2=r^2$; and (b) Euclid included the circle in his axioms. The actual naming aside, I will show that this particular arithmetic operation of *squaring*, as well as its inverse (*square root*), plays a crucial role in mathematics as well as in its applications.

Interestingly, unlike the idea of line or linearity, which is omnipresent among the various axioms of mathematics, squares and/or square roots are not. In fact, I am not even sure if they ever appear among the axioms of mathematics. Be that as it may, these two, the square and the square root, definitely appear among the postulates of physics.

It is almost impossible to find a fundamental law of nature that is not entirely based on linear and quadratic formulas (and their inverses). Indeed, a quick glance through science textbooks reveals Ampère's law, Lorentz's force, Coulomb's law, Newton's law, the law of, the special theory of relativity, as well as many others, and the (vast) majority of them are described using mathematical formulas that contains only linear or quadratic functions (and their inverses). Here are a few, just to prove the point:

$$m = m_0\sqrt{1-\frac{v^2}{c^2}} \qquad E = \frac{mv^2}{2} \qquad F_c = \frac{mv^2}{r} \qquad E = mc^2 \qquad T = 2\pi\sqrt{\frac{m}{k}}$$

$$F = am \qquad F = c\frac{m}{r^2} \qquad B = c\frac{I}{r} \qquad F = qE + qvB \qquad F = c\frac{q_1 q_2}{r^2}$$

$$\frac{V_1}{T_1} = \frac{V_2}{T_2} \qquad PV = \frac{mnc2}{3} \qquad \alpha = \sqrt{\frac{2KT}{M}} \qquad PV = nRT.$$

What is missing in the above formulas are $\log(x)$ and $\sin(x)$, and $\tan(x)$ and $\cosh(x)$. There are no complicated rational functions of cubic roots, and even slightly more complex polynomials are missing. Anything that is more complicated than r^2 is missing. Obviously, not all laws of nature are of this type, but, surprisingly, many are, and to be fair, many of those that are not expressed via these simple terms are often derived (and often via *linear* differential equations) from formulas that use these simple terms.

It is worth pondering on this for a moment or two. Mathematics contains myriad, countless in fact, formulas that are much more complicated (and frankly more interesting) than the expressions listed above. The law of gravity could have had a log in there, or a sin, or a hyperbolic function or two. But no, it has only linear terms and quadratic terms. The same is true for Coulomb's law as well as Einstein's relativistic corrections.

This fact is exceptionally useful to us humans, but, at the same time, somewhat unnerving. Why would the creator of our universe use these particular laws of nature? Had he made them just slightly (even a tiny bit) more complicated we would have had an insurmountable "mountain to climb." Had he used r^3 and not r^2 for the law of gravity (and why not, we live in a three-dimensional space), we would have had significantly more complicated theory and many more unsolved equations to deal with. (Imagine the n-body problem in that scenario.)

But he did not. The creator, just like the ancient Greek geometers, decided to stick with linear and quadratic equations, with lines and circle. And thank God for that.

That was physics and the square, and here comes mathematics and its infatuation with this operation.

- *Mathematical appeal.* As an arithmetic operation, squaring is a very pleasant one. Take this expression, for example: $|a+b|^2 + |a-b|^2$. Even a grade-school child could simplify it and produce its equivalent: $2a^2 + 2b^2$. Now try a slightly different expression: $|a+b|^3 + |a-b|^3$. Indeed, this tiny little alternation ($2 \to 3$) completely shatters our world. We have mathematically intractable formula.
- *Pythagoras.* To start with, this formula ($a^2 + b^2 = c^2$) is all about squares, and although it computes the length of a hypotenuse, its true significance comes from the fact that it actually computes *distance*; the shortest distance between two points. Surprisingly, this formula works for any dimension and it is always expressed with squares. This alone now explains so many r^2 (and not r^3) within the laws of physics.
- *L_2 space.* (again). It turns out that one can extend the idea of distance to general (functional) vector spaces, but it is only the L_2 space that allows for Pythagoras's Theorem. And the key ingredients are the squares that appear in the formula for distance. The number "2" seen within the very name L_2 space refers to the *squaring* operation.
- *Bell curve.* At first glance it seems impossible to connect the arithmetic operation of *squaring* with the coin toss problem, but in fact, this operation plays a crucial role. The bell curve appears via the Central Limit Theorem, but this appearance is predicated on one very important requirement: a finite second moment. In other words, EX^2 must be finite (i.e., the variance must be finite). This is important. Technically, this theorem could have called for a finite third moment, or one-point first moment or an exponential moment. But no! Only a second moment would do. Only the behavior of X^2 is of interest to the oracle of mathematics. (And do not ask why, for I have spent too many hours chasing this rabbit down its hole.)
- *Differential equations.* We have already commented on how lucky we were in this regard: The only differential equations we humans can solve comfortably are often the exact ones that follow from the laws of nature. We are talking about linear differential equations of the second order. The "second order" description is where the squaring operation comes alive. Indeed, with a simple calculus-type trick ($f=e^{rx}$), these equations transform into a *quadratic* equation.

Conclusion

There you have it, some ideas and some thoughts about this wonderfully complex and intriguing structure of mathematics. I hope the reader found it interesting, and maybe even has some ideas of her own. But I must warn you. Mathematicians are one of the most conservative and most orthodox people you will ever meet. Which is not that surprising – we follow a 3,000-year-old cult. Thus, tread carefully.

Appendix

The appendix contains several proofs and technical details that have been omitted in the main text for the sake of storytelling. They are here for the completeness reasons, but also because they are pretty and elegant. Besides the proofs the appendix offers brief discussions, expending on some of the ideas presented in the main text. In particular to: energy collapse of bouncing balls universe, and Brownian motion and the Newton laws.

Irrationality of $\sqrt{2}$

A simple application of Pythagoras's theorem implies that a diagonal of a unit square has length $\sqrt{2}$. Thus, clearly, this number exists. However, the Greeks were under the impression that all numbers should be expressible as fractions, which actually seems reasonable (try 99/70). So, let us see if this is possible. Let us assume that indeed $\sqrt{2} = n/m$ and that the fraction n/m is reduced (no common factors). This last comment plays an important role later on, so let us clarify: if there exists a fraction that equals $\sqrt{2}$ then we should be able to reduce it. Thus, we can assume that the form is already reduced.

Next, by squaring, $n^2 = 2m^2$, and we can conclude that n^2 is an even number. This, in turn implies that n must be even, since any odd number squared is odd (indeed $(2k+1)^2 = 4k^2 + 4k + 1$ is an odd number). And, since n is even, it means that $n = 2k$ for some integer k, which combined with $n^2 = 2m^2$ implies that $2k^2 = m^2$. But now we have that m^2 is even, and using the previous argument, we conclude that m must be even as well. Thus, both n and m are divisible by 2, which contradicts our assumption that n and m are integers without common divisors.

Discussion: This seemingly trivial little proof, an observation actually, shook the world of mathematics, and even now, some three thousand years later, we feel the tremor. The very existence of "the 13th axiom" of real numbers, the one that we need in order to have calculus, is a direct consequence of this "little proof." This result is the one that created different infinities (e.g., uncountable infinity) and, consequently, many mathematical paradoxes and bizarre theorems (and new theories needed to fix them). This result, coupled with Planck's quanta, creates an uneasy coexistence between quantum mechanics and mathematics. Indeed, quantum mechanics implies that many (if not all) variables used in physics must be fractions, but mathematical equations offer solutions that are typically irrational. So, which is it?

Computation Behind Planck's Story

The mathematics behind Planck's story offers a nice way to independently motivate many of the techniques learned in calculus sequence, things like Riemann sums, geometric series, L'Hospital's rule, function graphing, etc. One typically presents this tale in two steps, starting with the big picture and leaving the tricky math for later. Here is the "tricky" part.

Obviously Correct Formula

To refresh our memory: Planck's formula described the experimental results better than the other two formulas, but it could not be justified theoretically. Thus, before we proceed, it would make sense to compare the three formulas from a mathematical point of view.

Rayleigh-Jeans	*Wilhelm Wien*	*Max Planck*
$E(v)=\frac{8\pi v^2}{c^3}kT$	$E(v)=\frac{8\pi v^2}{c^3}\frac{k\beta v}{e^{\beta v/T}}$	$E(v)=\frac{8\pi v^2}{c^3}\frac{k\beta v}{(e^{\beta v/T}-1)}$.

Some simplification is in order. The only variable here is v, and the rest are constant, so in order to make the whole process more approachable we will factor out (i.e., erase) the constant $8\pi/c^3$ and replace v with the more familiar x. The three functions will be called $f(x)$, $g(x)$, and $h(x)$, respectively. With this little book-keeping done, the above formulas reduce to:

$$f(x)=kTx^2 \quad g(x)=\frac{k\beta}{e^{\beta x/T}}x^3 \quad h(x)=\frac{k\beta}{e^{\beta x/T}-1}x^3.$$

At first glance, the functions look very different, but once we plot them, we discover their similarities. All constants are set to one, and the functions are plotted on same chart:

Comments:

- The Rayleigh-Jeans formula ($f(x)$) obviously fails for large x's (i.e., ultraviolet catastrophe).
- The Rayleigh-Jeans formula and Planck's formula ($f(x)$ and $h(x)$) are nearly identical for small x's (circle on the left)
- Wien's formula ($g(x)$) fails for small x's (circle on the left)
- Wien's formula and Planck's formula ($g(x)$ and $h(x)$) are near identical for large x's (circle on the right)

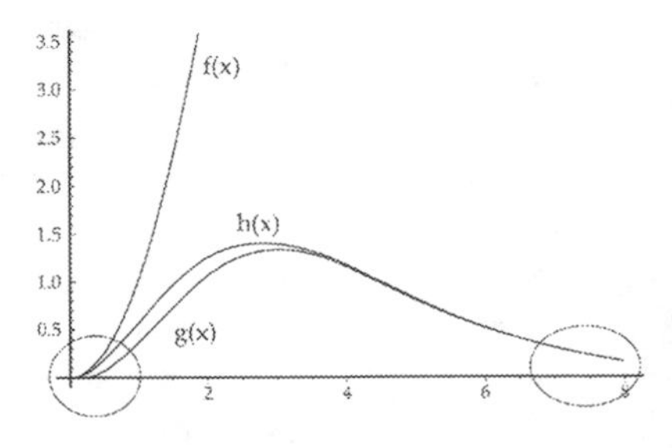

This visual inspection and intuition can be easily confirmed using limits. Indeed, regardless of the choice of constants we have the following:

$$\lim_{x\to\infty}\frac{g(x)}{h(x)}=\lim_{x\to\infty}\frac{e^{\beta x/T}-1}{e^{\beta x/T}}=1-\lim_{x\to\infty}\frac{1}{e^{\beta x/T}}=1.$$

$$\lim_{x\to 0}\frac{h(x)}{f(x)}=\lim_{x\to 0}\frac{\frac{k\beta}{e^{\beta x/T}-1}x^3}{kTx^2}=\lim_{x\to 0}\frac{\beta x}{Te^{\beta x/T}-1}\text{ (L'Hospital)}=\lim_{x\to 0}\frac{1}{e^{\beta x/T}}=1.$$

Justifying "−1" in Planck's formula

Instead of using calculus formulas, Planck applied a direct computation, via a Riemann sum. In other words, he used the following approach:

$$\int_0^\infty f(x)dx=\lim_{\delta\to 0}\sum_{i=0}^\infty \delta f(i\delta).$$

Applied to Boltzmann's formula (after replacing variable E with more familiar x) this integration formula yields the following:

$$\int_0^\infty e^{\frac{-x}{kT}}dx=\lim_{\delta\to 0}\sum_{i=0}^\infty \delta e^{\frac{-i\delta}{kT}}\text{ and }\int_0^\infty xe^{\frac{-x}{kT}}dx=\lim_{\delta\to 0}\sum_{i=0}^\infty i\delta^2 e^{\frac{-i\delta}{kT}}.$$

Consequently,

$$\bar{E}=\frac{\int_0^\infty xe^{\frac{-x}{kT}}dx}{\int_0^\infty e^{\frac{-x}{kT}}dx}=\lim_{\delta\to 0}\frac{\sum_{i=0}^\infty \delta^2 ie^{\frac{-i\delta}{kT}}}{\sum_{i=0}^\infty \delta e^{\frac{-i\delta}{kT}}}.$$

We simplify the notation with $W=e^{\frac{-\delta}{kT}}$, and observe that now $e^{\frac{-i\delta}{kT}}=W^i$. Therefore, the above expression contains two series for which we have a closed form solution.

$$\sum_{i=0}^\infty W^i=\frac{1}{1-W}\text{ and }\sum_{i=0}^\infty iW^i=\frac{W}{(1-W)^2}.$$

After plugging these two back into the limit, and with little algebra we get:

$$\bar{E}=\lim_{\delta\to 0}\frac{\delta}{W^{-1}-1}\text{ (since }=e^{\frac{-\delta}{kT}}\text{)}=\lim_{\delta\to 0}\frac{\delta}{e^{\frac{\delta}{kT}}-1}.$$

If we were to compute this limit (as calculus would suggest) we would apply L'Hospital's rule and got the following:

$$\lim_{\delta\to 0}\frac{\delta}{e^{\frac{\delta}{kT}}-1}=\lim_{\delta\to 0}\frac{1}{(kT)^{-1}e^{\frac{\delta}{kT}}}=kT,$$

which yields the Rayleigh-Jean's formula. However, if we introduce a new law of physics (minimum amount of energy - *quanta*), then we do not have $\rightarrow 0$, and one can recover Planck's correct formula.

Bouncing Balls and Energy Collapse

In chapter on bouncing balls, we described the issue of vanishing energy and mentioned the random walk as a model. So, let us refresh our memory. We start our universe with m balls each with its own mass and velocity, thus we have total initial energy E_0. Although the law of conservation of energy would suggest otherwise, this initial energy is not preserved. After each collision, as we have described in the earlier chapter, the mathematics and the laws of physics are in a disagreement. On our computer we cannot keep all the decimal points of resulting velocities, and, consequently, after each collision, we are either adding or subtracting a minute amount of energy. Let us call this loss or gain of energy by B_k (Note: this variable does not present kth ball but kth collision).

Since we do not know what was the truncated digit (did we round up or down?), we assume that it is random, and that it has a 50–50 chance of being positive or negative. Example: if velocity after collision is 1/3, the computer can keep only 0.33333333, thus we are *losing energy*, but if velocity is 2/3 we keep 0.66666667 and consequently we are *adding energy*. Therefore, total energy (after n collisions) is:

$$E_n = E_0 + \sum_{k=1}^{n} B_k ,$$

and since B_k is extremely small, random and equally likely to be positive or negative, the above construction behaves like a Brownian motion/random walk. The only difference is that this time we start at E_0 (and not at zero as usual).

A picture is worth a thousand words. Figure A.1 shows the energy fluctuations: E_n.

The inevitability of the energy collapse now follows trivially. Total energy E_n (like a Brownian particle) oscillates up and down through the time. And, although

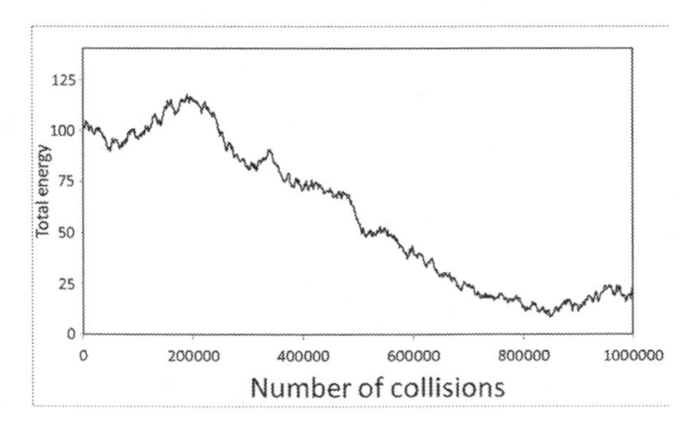

Figure A.1 Bouncing balls energy level.

it might take some time (millions of collisions), with mathematical certainty we can predict that Brownian motion will hit $E_n = 0$, level. Translated to our case, this means zero energy, which means zero velocities, which means that the bouncing balls are "frozen" on the computer screen.

But in reality, the situation is even worse. As the totally energy E_n randomly fluctuates (up and down) we observe an interesting phenomenon. During "up" periods, we see uptake in total energy E_n, and consequently the balls are moving faster (large kinetic energy). But having faster moving balls leads to more collisions and faster energy fluctuations. Consequently, this universe does not stay too long in this "up" period. However, during the "down" periods the opposite is true. We see less energy, slower balls, fewer collisions, and, consequently our universe slows down even before the total energy hits zero. Moreover, once the total energy is small enough, it stays small "forever." Indeed, the arcsine law dictates that the universe stays stuck in this (sluggish) state for an exceedingly long time.

Comment: If one implements computation with 16 (or 32) digits precision, this "slowdown" will not be apparent. In fact, it might take a while (maybe even centuries) before we observe any deceleration. But, no matter the precision of our computers, this universe will collapse. Mathematics is quite certain about that.

Uncertainty Principle and Matrix Algebra

Here we show only a polished outline of the mathematics behind the uncertainty principle. For two arbitrary Hermitian operators $A: H \to H$ and $B: H \to H$, and any element x of H such that ABx and BAx are both defined (so that in particular, Ax and Bx are also defined), then

$$\langle BAx \mid x \rangle = \langle Ax \mid Bx \rangle = \langle Bx \mid Ax \rangle^{*}.$$

In an inner product space the Cauchy-Schwarz inequality holds.

$$|\langle Bx \mid Ax \rangle|^2 \le \| Ax \|^2 \| Bx \|^2.$$

Rearranging this formula leads to

$$\| Ax \|^2 \| Bx \|^2 \ge |\langle Bx \mid Ax \rangle|^2 \ge |im(\langle Bx \mid Ax \rangle|^2 = \frac{1}{4}|2im(\langle Bx \mid Ax \rangle|^2$$

$$= \frac{1}{4}|\langle Bx \mid Ax \rangle - \langle Ax \mid Bx \rangle|^2 = \frac{1}{4}|\langle ABx \mid x \rangle - \langle BAx \mid x \rangle|^2 = \frac{1}{4}|\langle (AB - BA)x \mid x \rangle|^2.$$

Consequently, the following general form of the uncertainty principle, first pointed out by Howard Percy Robertson and (independently) by Erwin Schrödinger, holds:

$$\frac{1}{4}|\langle (AB - BA)x \mid x \rangle|^2 \le \| Ax \|^2 \| Bx \|^2.$$

Heisenberg's uncertainty relation is immediate: Suppose A and B are two observables which are identified as self-adjoint (and in particular symmetric) operators. If we define the "average" operator as $\langle X \rangle_{\varphi} = \langle \varphi \mid X\varphi \rangle$ and "Standard Error" operator as $\Delta_\varphi X = \sqrt{\langle X^2 \rangle_\varphi - \langle X \rangle_\varphi^2}$, and if we assume that both $B\,A\,\psi$ and $A\,B\,\psi$ are defined then the above inequality yields

$$\Delta_\varphi A \Delta_\varphi B \geq \frac{1}{2}\left|\langle AB - BA \rangle_\varphi\right|.$$

People skilled in art will quickly recognize the simple mathematical mechanism that governs this principle. Namely, $AB \neq BA$ and consequently the right-hand side is strictly greater than zero. *As a result, the "error" is greater than a fixed constant!* One of the most bizarre laws of physics is there because matrix multiplication is not commutative!

Measure Theory Problem

Let us start with "reasonable" axioms designed to capture the idea of measure of a set (or a length of a subset of a real line).

For any $A \subset R$ we define a function: $\lambda(A)$, a generalized "length" such that

Axiom 1. $\lambda(A) \geq 0$

Axiom 2. $\lambda(\{\}) = 0$

Axiom 3. For sequence $\{A_i\}_{i=1}^{\infty}$ *where* $i \neq j \Rightarrow A_i \cap A_j = \{\}$ *we have the following* $\lambda(\cup_{i=1}^{\infty} A_i) = \sum_{i=1}^{\infty} \lambda(A_i)$

Axiom 4. For any real number a, $\lambda(A+a) = \lambda(A)$ *and* $\lambda(-A) = \lambda(A)$

Axiom 5. For any $a \leq b$ *and* $A = [a,b] \Rightarrow \lambda(A) = b - a$

Theorem 1. It is impossible to construct function $\lambda(A)$ so that Axioms A1–A5 are satisfied.

The proof follows easily from the following lemma.

Lemma 1. There exists a set $A \subset [-1,1]$ and a sequence of numbers $\{q_i\}_{i=1}^{\infty}$ such that sets $A_i = A + q_i$ are disjoint and such that $[-1,1] \subset \cup_{i=1}^{\infty} A_i \subset [-3,3]$.

It is easy to show that if Lemma 1 is true, we have a contradiction. First, from the above axioms it is easy to show that $B \subset D \Rightarrow \lambda(B) \leq \lambda(D)$. Next, we observe that since $\{A_i\}_{i=1}^{\infty}$ are disjoint we have (by A3) that $\lambda(\cup_{i=1}^{\infty} A_i) = \sum_{i=1}^{\infty} \lambda(A_i)$. However, by A4 we have that $\lambda(A_i) = \lambda(A) = \delta$, for all *i*'s. Now if $\delta = 0$, then (by A3) $\lambda(\cup_{i=1}^{\infty} A_i) = 0$, which is in contradiction with the fact that $2 = \lambda([-1,1]) \leq \lambda(\cup_{i=1}^{\infty} A_i) = 0$. On the other hand, if $\delta > 0$ then (by A3) $\lambda(\cup_{i=1}^{\infty} A_i) = \infty$, which is in contradiction with the fact that $\infty = \lambda(\cup_{i=1}^{\infty} A_i) \leq \lambda[-3,3] = 6$. This proves Theorem 1.

Proof for Lemma 1.

For any x and y in $[-1,1]$, we define relation "$\equiv$" by saying $x \equiv y$ if $x - y = p$ for some rational number p. This relation defines an "equivalency" (it is easy to check that $x \equiv x$, and $x \equiv y \Rightarrow y \equiv x$, and finally $x \equiv y \,\&\, y \equiv z \Rightarrow x \equiv z$). Thus, the set $[-1,1]$ can be partitioned with the equivalency class.

(This construction is a bit tricky, so an example will help: take $x = \sqrt{2}/2$ and let B_x contain all $y \in [-1,1]$ for which $y - x$ is a rational number. Think of all rational "shifts" of x. One can do the same with $z=\sqrt{2}/3$, and construct a set B_{xz} analogously. Clearly $B_{xz} \cap B_x = \{\}$ since $\sqrt{2}/2-\sqrt{2}/3$ is not rational. Now, set theory guarantees that all the sets that could be created in this manner are disjoint and their union is equal to $[-1,1]$.)

Next, we define a set A by taking one (and only one) element from each of this partition sets. Axiom of choice guarantees that we can have such a set A. Next, by enumerating rational numbers in $[-2,2]$, we can construct a sequence $\{q_i\}_{i=1}^{\infty}$ where $i \neq j \Rightarrow q_i \neq q_j$ and $\{q_i\}_{i=1}^{\infty} = Q \cap [-2,2]$. Finally, we define $A_i = A + q_i$.

First, we show that these sets are disjoint (i.e., $i \neq j \Rightarrow A_i \cap A_j = \{\}$).

Proof. Suppose not. Thus, there exists a and b in A and such that $a+q_i = b+q_j$, which implies that $a-b = q_j - q_i = p$. Since p is rational, we conclude that both a and b are in the same partition set. However, A was specifically chosen by picking only one representative from each partition set, thus $a = b$, which implies $q_i = q_j$ which is a contradiction.

Second, we show that $[-1,1] \subset \bigcup_{i=1}^{\infty} A_i$.

Proof: Let $x \in [-1,1]$ then x must belong to one of the partitioning sets and therefore there must exist one (and only one) $a \in A$ from the very same partition. Since both x and a are in the same partition set, we have that $p = x-a$ is rational. Moreover, since both x and a are in $[-1,1]$, we conclude that p is in $[-2,2]$. Thus, there exists $q_i = p$ which in turn means that $x = a + q_i \in A_i$ which concludes the proof.

Third, we need to show that $\bigcup_{i=1}^{\infty} A_i \subset [-3,3]$.

Proof: Trivial. Since $A \subset [-1,1]$ and $\{q_i\}_{i=1}^{\infty} \subset [-2,2]$.

This proves Lemma 1 and Theorem 1.

Brownian Motion and Newton's Laws

We mentioned this idea in the book but it pays to revisit and dig a bit deeper. To start with, we should recall the notion of *continuum* and its relationship to a moving object. *Continuum* was one of the fundamental concepts behind Newton's ideas, and for a good reason. In nutshell it says the following: an abstract mathematical formula $f(t)$, if it represents a *continuous* function then it is naturally associated to a movement of an object. And by now we are all familiar with the narrative: $f(t)$ represents a position of the object at the time t.

Applied to our case, since the mathematical model for Brownian motion $W(t) = (X(t), Y(t))$ is a continuous function, we can say that $W(t)$ naturally represents a movement of an object. That would be the theory. The experiment is even easier to argue, since after all, the whole story started with an observation of a movement of an actual object (pollen particles).

That was the easy part. Here comes the problem: this motion, this continuous function, does not agree with Newton's laws, indeed.

1) The second law stipulates that an object should accelerate in a direction of the resultant force. However, since at any time, the particle described by the function $W(t)$ does not move in any preferred direction, we must conclude that at any moment in time, the resultant force is zero.

2) Newton's first law states that an object on whom the resultant force is zero must either stay still or move along a straight line. But $W(t)$ is obviously not a straight line, which now implies that there must exist an outside force.

So, which is it? Do we have force acting on this particle or not?

One might try to dismiss this argument and declare that no particle actually follows the path of Brownian motion, and that $W(t)$ is just a mathematical abstraction designed to approximate reality. Well, this is obviously true with Brownian motion, but then again, it is also true with essentially any other particle motion. They are all described by mathematical formulas that are in their essence just abstractions designed to approximate reality. For example, under close inspection Earth does not follow an elliptical trajectory; Jupiter and other planets disturb its path.

But even if one accepts this argument that actual pollen particle does not really follow the path of $W(t)$, we still have an issue to address. Namely, $W(t)$ is very much real and very much continuous. Consequently, it is associated with a moving object (pollen particle or not).

At first glance we have a peculiar situation: an abstract mathematical construction that suggests some bizarre corrections to the laws of physics. However, through history, this "peculiar" scenario has been played out many a time. This is how we got *quanta* and this is how we got *antimatter* among many other things. (Side note: It seems that Dirac first "discovered" antimatter after "peculiar" math: $E^2 = m^2c^4 \rightarrow E = \pm mc^2$ suggested "negative" energy.) So, who is to say that adding Brownian motion to the inertia postulate would not yield some interesting physics?

One way to explore this idea is to adopt Einstein's fantasy. Instead of visualizing what would happen if we travel on an electromagnetic wave, let us imagine ourselves as passengers in a space shuttle that moves along the path of Brownian motion. A few easy observations:

a. *You do not feel any force,* since if you did, you would have been able to detect the direction of your movement (and this you cannot do).
b. *You move, but have no idea where.* Indeed, if you stop this journey after say $t = 100$ seconds, you will be far away from the starting point, but with no control over the direction of this journey.
c. *On average you do not move at all.* If you repeat this journey many times, you will equally likely move left and right (up and down), and all this cancels out, leaving you at the same spot you started (on average).

Some tricky questions:

d. *If you look out the window, what do you see?* Since at any moment in time your shuttle is moving in "all directions," the view "in front of you" should consist of all possible, 360° direction views (I guess).
e. *Does Einstein's first postulate still apply?* In other words, can you perform any experiment inside this shuttle to detect if you are moving or staying still?
f. *Does the inclusion of Brownian motion as a third possible motion under inertia contradicts any laws of physics?*

And as a final teaser, *True or false*: every continuous function $f(t)$ is associated with a moving object?

If yes, then the path of the Wiener process must be included as one of the movements allowed under the inertia postulate.
If no, then we would need a guideline. Which continuous functions are associated with a moving particle and which are not?

On the Quantum Conundrum

We have talked about this in the main body of the book as well as in the appendix. To refresh our memory: after particle collisions, due to the quantum postulate, we must allocate energy (velocities) to various particles in a form of kq (where k is an integer and q a fixed constant). Mathematically, this is typically impossible to do since the equations offer solutions in the form of a real number (often irrational). Consequently, we (i.e., the universe) must round off the velocities in order to satisfy the quantum postulate.

Since the universe does not have any control over the round-off digit, after every collision it must either randomly add or subtract a minute amount of energy. Unfortunately, this fluctuation does not completely cancel out, and (as we have shown earlier), the total amount of energy behaves very much like Brownian motion. Consequently, the total energy of the system (in this case our universe) will eventually collapse to zero. There does not seem to be any way out of this, unless I am misreading the laws of physics.

Now, to be fair, this random oscillation per collision is indeed minute ($q \approx 10^{-34}$). On top of that, the mathematics of random walk implies that total energy oscillation is on the order of $q\sqrt{n}$; where n is the number of collisions. Thus, it seems that we would need an enormous number of collisions (enormous n) before the effect is detectable. In other words, it might take a very long time.

On the other hand, there are many (many) such collisions in our universe. Tony Padilla (of the University of Nottingham) had recently provided an estimate of circa 10^{80} such particles. If we factor in that most of them are colliding (interacting) with other particles and that they do so with an astonishing frequency (millions per second), and that they have been doing so for billions of years, then, well, we might have something. It all adds up, doesn't it? Is it enough to be noticeable? Who knows?

But it gets worse, since we do not need any collisions or fancy constructions to create a contradiction. Just a plain simple object with mass m in a gravitational field is enough. Namely, the laws of physics insist that

a. this object's total energy (potential E_p plus kinetic E_k) must stay constant.
b. both kinetic and potential energy must satisfy quantum postulate $E_p = iq$ and $E_k = jq$ for some integers i and j.

Trivial computation now reveals that $C = E_p + E_k \rightarrow i + j = C/q$, which is *impossible* to achieve (for the vast majority of integers i and j). Thus, the universe must "round off" the energy and assign i's and j's so that the following holds:

$$i + j = C/q + \varepsilon \text{ for some small } \varepsilon.$$

How often does universe do this? Well, let us see. Our particle cannot travel at any speed. It has kinetic energy, and since quantum postulate insists that energy must be of the form kq (where k is an integer), we have the following

$$E = kq = v_k^2 m / 2, \rightarrow v_k = \sqrt{2kq / m} \text{ for } k = 1, 2, 3, ...$$

Here comes the trouble: as our particle accelerates in this gravitational field, its velocity cannot continuously increase, it has to jump from v_k to v_{k+1}, and each time this happens, the universe must re-do the above computation and commit another of these round-off errors small ε. But these jumps happen extraordinarily often! Every time we have even a minute acceleration. In fact, these rounding errors occur so often that the total error (due to random walk) would collapse the total energy of this particle within a few hours (if not minutes).

The reader is welcome to simulate this movement on her computer. Just make the constant q bigger (say $q = 0.001$), implement Newton's laws and quantum the postulate (i.e., $v_k = \sqrt{2kq / m}$), keep the force F constant to make it simpler, and watch. But be careful: the quantum condition on potential energy now forces that a position of this particle cannot be arbitrary. When all said and done, when this is properly implemented, the total energy $C = E_p + E_k$ collapses very quickly.

The obvious issue is the discrepancy between Newton's laws, which deal with continuous functions, and the quantum postulate which insists on discrete solutions. But these two, discrete and continuous mathematics, they do not "like" each other very much. The vast majority of continuous equations do not have discrete solutions. That is not my theory, but a simple mathematical fact. How does our universe solve this conundrum? I have no idea. Does it "memorize" the true velocities and energies, but "displays" only discrete (rounded) versions? Who knows?

Recommended Reading

Introducing the Mystery

About Newton, Halley, and the paths of heavenly bodies, see James Gleick, *Isaac Newton*, Vintage Books, 2004. More on the remarkable history of imaginary numbers can be found in Paul Nahin's *An Imaginary Tale: The Story of the Square Root of −1*, Princeton University Press, 1998. As far as Diophantine equations are concerned, probably the best source to start with would be *Diophantus and Diophantine Equations* by Izabella Bashmakova, Mathematical Association of America, 1997. The original is in Russian but there is an English translation. There are many good sources on Greek and Renaissance mathematics; one of my favorites is Luke Hodgkin's *Greeks and Origins – A History of Mathematics: From Mesopotamia to Modernity*, Oxford University Press, 2005. Eugene Wigner's paper "The Unreasonable Effectiveness of Mathematics in the Natural Sciences" can be found in *Communications in Pure and Applied Mathematics*, vol. 13, no. 1 (February 1960). Richard Hamming offers a follow-up to Wigner's article titled "The Unreasonable Effectiveness of Mathematics," in *The American Mathematical Monthly*, vol. 87, no. 2 (February 1980). Max Tegmark's book *Our Mathematical Universe: My Quest for the Ultimate Nature of Reality*, Knopf, 2014 – as well as Graham Farmelo's book *The Universe Speaks in Numbers: How Modern Math Reveals Nature's Deepest Secrets*, Basic Books, 2019 – are closely related to the task I am undertaking.

Among the mathematical historians/philosophers/linguists/ psychologists, here are few whose work relates to the task at hand: *Where Mathematics Comes From* by George Lakoff and Rafael Nunez, Basic Books, 1992; *Trick or Truth? The Mysterious Connection between Physics and Mathematics*, ed. by Anthony Aguirre, Brendan Foster, and Zeeya Merali, Springer, 2016; *The Applicability of Mathematics as a Philosophical Problem* by Mark Steiner, Harvard University Press, 1998; *Being and Event* by Alain Badio, Continuum International Publishing, 2006.

The influence of G. H. Hardy's essay *A Mathematician's Apology*, Cambridge University Press, 1940, on generations of mathematicians is hard to overstate. Hardy's book embodies much of the spirit presented in my book. Richard

Feynman's autobiography *Surely You're Joking, Mr. Feynman!*, W. W. Norton & Company, 2010, is another must-read, particularly as it offers Feynman's witty and perceptive intuition regarding the math–physics interplay. On amicable numbers and the golden ratio one can read D. Wells, *Dictionary of Curious and Interesting Numbers*, Penguin Books, 1998. On Greek philosophy there is an abundance of material; one of my favorites is S. Cohen, P. Curd, and C. D. C. Reeve, *Readings in Ancient Greek Philosophy: From Thales to Aristotle*, Hackett Publishing, 2016. About pi and, in particular, its early history, see *The Pythagorean Theorem: A 4,000-Year History* by Eli Maor, Princeton University Press, 2007.

On Classical Mathematics

On calendars, as well as the formation of the first civilizations, more can be found in David Ewing Duncan, *Calendar: Humanity's Epic Struggle to Determine a True and Accurate Year*, Harper Perennial, 1999; Michael Wood, *In Search of the First Civilizations*, BBC Books, 2007; and Max Weber, *The Agrarian Sociology of Ancient Civilizations*, Verso, 2013. On the origins of conic sections and the role of Menaechmus, see Roger Cooke, *The Euclidean Synthesis, the History of Mathematics: A Brief Course*, Wiley, 2011. About Apollonius of Perga, a direct, yet slightly obscure source is Ivor Bulmer-Thomas's *Selections Illustrating the History of Greek Mathematics*, Forgotten Books, 2018. For the fascinating subject of Arab and medieval Islamic mathematics, there are many sources, but unfortunately mainly in the form of lecture notes or chapters in books. See, for example, *The History of Mathematics: A Brief Course* by Roger L. Cooke, Wiley, 2011, as well as *A History of Mathematics* by Carl B. Boyer and Uta C. Merzbach, Wiley, 2011. A more approachable book focused on this subject would be very welcome. There are many books on Fourier and his analysis, but I have my favorites. The book *Who Is Fourier? A Mathematical Adventure* by Transnational College of LEX, Language Research Foundation, 1995, is unusually approachable and yet very informative. So is T. W. Körner's *Fourier Analysis*, Cambridge University Press, 1998, in which the reader will learn more than she bargained for.

To read more about logic and logical circuits, see *The Logician and the Engineer: How George Boole and Claude Shannon Created the Information* by Paul J. Nahin: Princeton University Press, 2013; and *The Life and Work of George Boole: A Prelude to the Digital Age* by Desmond MacHale, Cork University Press, 2014.

Weierstrass's nowhere differential continuous function, and as well as Peano's space-filling curve, are well referenced and easy to find. In particular, about their fractal structure, see *The Fractal Geometry of Nature* by Benoit Mandelbrot, Echo Point Books & Media, LLC, 1982.

I try to avoid referencing original papers here, but Robert Brown's original work is a masterpiece and I had to make an exception. The paper is informative and easy to read, and it shows how real science should be conducted. It appeared in *Philosophical Magazine*, 1828, and was titled: "A Brief Account of Microscopical Observations Made in the Months of June, July, and August 1827, on the Particles Contained in the Pollen of Plants; and on the General Existence of Active Molecules in Organic and Inorganic Bodies." About Bechalier's and his thesis one can read *Bechalier and His Times: A Conversation with Bernard Bru* by Murad Taqqu, Finance Stochast, 2001. On Brown-Bachelier-Einstein-Wiener saga, there is a wealth of references. *Introduction to Stochastic Differential Equations with Applications to Modelling in Biology and Finance* by Carlos A. Braumann, Wiley Online Library, 2019, is one suited for more general audience. On Galileo and his impact there are abundance of references, one of my favorites is a lesser-known book: *Galileo Galilei: Father of Modern Science* by Rachel Hilliam, Rosen Publishing Group, 2004. As for Lavoisier, a father of chemistry, see *Lavoisier in the Year One: The Birth of a New Science in an Age of Revolution* by Madison Smartt Bell, W. W. Norton & Company, 2006.

On Modern Physics

The whole of Chapter 3 (and in particular Section 3.1 on Planck and Quanta) is very much influenced by the book: *What is Quantum Mechanics? A Physics Adventure,* by Transnational College of LEX, Language Research Foundation, 1995. Another book that played a role is *Quantum Theory for Mathematicians* by Brian C. Hall, Springer-Verlag, 2013. On black-body radiation and Max Planck, one can see *Planck: Driven by Vision, Broken by War,* by Brandon R. Brown, Oxford University Press, 2015. Einstein and his motivations, as well as the anecdotes behind his work, have been well documented and there is a wealth of books available.

On uncertainty and Heisenberg one of my favorite books is *Uncertainty: Einstein, Heisenberg, Bohr, and the Struggle for the Soul of Science* by David Lindley, Anchor, 2008. Nevertheless, the aforementioned book *What is Quantum Mechanics?* offers a better account of the actual math/physics behind this phenomenon. Regarding the claim about matrix theory and the Uncertainty Principle I offer a quote: "Heisenberg's groundbreaking paper of 1925 neither uses nor even mentions matrices." More can be found in a Wikipedia article: "Heisenberg's Entryway to Matrix Mechanics" https://en.wikipedia.org/wiki/Heisenberg%27s_entryway_to_matrix_mechanics. Also see *Entanglement* by Amir Aczel, Plume, 2003; *In Search of Schrödinger's Cat: Quantum Physics and Reality* by John Gribbin, Bantam, 1984; and *Einstein's Universe: Gravity at Work and Play* by A. Zee, Oxford University Press, 1989.

On Computer Games

As expected, there are numerous books written on chess. One that addresses the history as well as aspects of the game discussed here is *The Immortal Game: A History of Chess* by David Shenk, Anchor, 2007. The official rules of the International Chess Foundation can be found on their website: https://handbook.fide.com/. Coding a basic chess game on computer is not that hard. See www.freecodecamp.org/news/simple-chess-ai-step-by-step-1d55a9266977/ and for checkmate in 268 moves see: https://gothicchess.info/programs_03.shtml.

About Convey see *Genius at Play: The Curious Mind of John Horton Conway* by Siobhan Roberts, Bloomsbury USA, 2015; more on the Game of Life can be found on

Wikipedia: https://en.wikipedia.org/wiki/Conway27s_Game_of_Life. For more on game designs and their programming see *The Art of Game Design: A Book of Lenses* by Jesse Schell, Morgan Kaufmann Publishers, 2008, and *Programming Game AI by Example* by Mat Buckland Jones, Bartlett Learning, 2005. On round-off error and the Patriot Missile glitch see *The New York Times*, June 6, 1991: www.nytimes.com/1991/06/06/world/us-details-flaw-in-patriot-missile.html.

On Mathematical Logic

For more on stalemate and chess in general see *A History of Chess* by Murray, H. J. R., Skyhorse, 2015. Regarding the saga behind Euclid's fifth postulate, see *The Fifth Postulate: How Unraveling a Two Thousand Year Old Mystery Unraveled the Universe* by Jason Socrates Bardi, Wiley, 2008. As for E. Beltrami and his work see: www.britannica.com/biography/Eugenio-Beltrami.

Measure theory is a tough subject to learn as well as to teach. If I were to choose a person who could best present this subject to a wider audience it would be David Pollard; his book is *A User's Guide to Measure Theoretic Probability*, Cambridge University Press, 2002. A book on the Axiom of Choice that nicely complements the ideas presented here is *The Axiom of Choice* by Thomas J. Jech, Dover Publications, 2008.

About anthropological and psychological origins of integer numbers see *Numbers and the Making of Us: Counting and the Course of Human Cultures* by Caleb Everett, Harvard University Press, 2017, as well as *PI in the Sky: Counting, Thinking, and Being* by John D. Barrow, Back Bay Books, 1992. About Bertrand Russell's work see, from the man himself, *The Principles of Mathematics*, W. W. Norton & Company, 1996. More approachable text can be found at https://plato.stanford.edu/entries/russell-paradox/. See also *Introduction to Mathematical Philosophy* by Bertrand Russell, Dover Publications, 1993. About the founder of set theory see *Georg Cantor: His Mathematics and Philosophy of the Infinite* by Joseph Warren Dauben, Princeton University Press, 1990. On clever quotes see *Mathematical Maxims and Minims* by Nicholas J. Rose, Rome Press, 1988. On Incompleteness Theorem a more technical book is *Gödel's Proof* by Ernest Nagel, New York University Press, 2001, while very informative and approachable text can be found: at https://plato.stanford.edu/entries/goedel-incompleteness/.

On Postulates and Axioms

For more on the three Newton postulates see *Isaac Newton and the Laws of the Universe: Physical Science* by Jane Weir, Teacher Created Materials, 2007. Harmony of spheres as ancient Greek (Pythagoras) concept lingered through the Western civilization for millennia. Even Johannes Kepler published a book on the subject, *Harmonices Mundi* (1619). An excellent source on Fermat's Theorem is the book *Fermat's Last Theorem* by Simon Singh, Harper Press, 2012. Regarding Poincaré – his conjecture as well as Perelman's contributions – see, *The Poincaré Conjecture: In Search of the Shape of the Universe* by

Donal O'Shea, Walker Books, 2008. See also *Men of Mathematics* by E. T. Bell, Touchstone, 1998, and *Adventures of a Mathematician* by S. M. Ulam, Daniel Hirsch, William G. Mathews, Françoise Ulam, &, Jan Mycielski, University of California Press, 1991.

Introducing the Oracle

An excellent book that introduces the long-standing open (and not very useful) math problems is *Mage Merlin's Unsolved Mathematical Mysteries* by Satyan Devadoss and Matt Harvey, The MIT Press, 2021. Another very good book (which should not be judged by its cover) is *The Manga Guide to Cryptography* by Masaaki Mitani, Shinichi Sato, and Idero Hinoki, No Starch Press, 2018. Information on doubling the cubes and other impossible problems can be found in *Tales of Impossibility: The 2000-Year Quest to Solve the Mathematical Problems of Antiquity* by David S. Richeson, Princeton University Press, 2019.

About hyperbolic function see: www.britannica.com/science/hyperbolic-functions. For numbers and a near infinite number of (useless) theorems see *Numbers are Forever: Mathematical Facts and Curiosities* by Liz Strachan, Constable, 2013. An interesting take on an old problem can be found in *Uncle Petros and Goldbach's Conjecture* by Apostolos Doxiadis, Bloomsbury, 2001.

On Probability

The Poisson process is typically covered as a chapter within a book and it is hard to find an approachable manuscript that is devoted solely to this topic. An excellent (albeit technical) treatment can be found on the MIT online source: "Poisson Processes – MIT OpenCourseWare" https://ocw.mit.edu/courses/6-041-probabilistic-systems-analysis-and-applied-probability-fall-2010/resources/lecture-14-poisson-process-i/.

Brownian motion is another technical topic for which is hard to find easy-to-follow books. A technical but more approachable book (with a very good introductory chapter) is *Brownian Motion: An Introduction to Stochastic Processes* by René L. Schilling and Lothar Partzsch, De Gruyter, 2014.

Somewhat controversial, but nevertheless a good source, is the book *The Bell Curve: Intelligence and Class Structure in American Life* by Richard J. Herrnstein and Charles Murray, Free Press, 1996. For a more mathematical book on this topic see *The Life and Times of the Central Limit Theorem* by William J. Adams, American Mathematical Society/London Mathematical Society, 2009. About the foundations of probability theory see www.britannica.com/biography/Andrey-Nikolayevich-Kolmogorov.

Post Scriptum

Arguably one of the best books on Euclid and his original work is *The Thirteen Books of the Elements* by Thomas L. Heath, Dover Publications, 1956. To learn more about Decartes, his life, philosophy, and mathematics, see *Descartes' Bones: A Skeletal History of the Conflict between Faith and Reason* by Russell Shorto, Vintage, 2009. For the Aline story and message hidden in the digits of π, see *Contact* by Carl Sagan, Pocket Books, 1997. See also *The Road to Reality: A Complete Guide to the Laws of the Universe* by Roger Penrose, Vintage, 2007; *God and the New Physics* by Paul Davies, Simon & Schuster, 1984; *The Colossal Book of Mathematics: Classic Puzzles, Paradoxes, and Problems* by Martin Gardner, W. W. Norton & Company, 2001; *Mathematical Mysteries: The Beauty and Magic of Numbers* by Calvin C. Clawson, Basic Books, 1996; *The Story of Mathematics: From Creating the Pyramids to Exploring Infinity* by Anne Rooney, Arcturus Publishing, 2011; and *The Mathematical Century: The 30 Greatest Problems of the Last 100 Years* by Piergiorgio Odifreddi, Princeton University Press, 2000.

And finally my favorite two: *The Mathematical Universe: An Alphabetical Journey through the Great Proofs, Problems, and Personalities* by William Dunham, John Wiley & Sons, 1994, and *The Mathematical Experience* by Phillip J. Davis and Reuben Hersh, Houghton Mifflin, 1981.

Index

Printed in the United States
by Baker & Taylor Publisher Services